My Thoughts Exactly

BRYTON J. ZAAGMAN

Copyright © 2022 Bryton J. Zaagman
All rights reserved
First Edition

PAGE PUBLISHING, INC.
Conneaut Lake, PA

First originally published by Page Publishing 2022

ISBN 978-1-6624-3748-9 (pbk)
ISBN 978-1-6624-5398-4 (hc)
ISBN 978-1-6624-3749-6 (digital)

Printed in the United States of America

Preface

The time is 9:04 a.m. It is Monday, August 28, 2017. It is the moment I decided to take action. You see, I've always thought of myself as a good writer, whether it be a story of fiction, a fact-based essay for a college class, or a biographic introduction to a résumé for the world of business. I have always scored well on essays and research papers for school. I even had a university professor compliment me on a paper I wrote about the exponential expansion of the Universe. I also four-pointed an English class at Grand Rapids Community College in Michigan. With the wealth of knowledge, I have obtained over my short but sweet twenty-six and a half years of life on Earth, I figured it was about damn time I turn my thoughts into reality. The inception of this came after a deep dive into the so-called Pandora's box. Fueled by concentrated tetrahydrocannabinol, dopamine stimulants, caffeine, nicotine, alcohol, and mania, I was dizzyingly lost in thought as day and night became an entire week without sleep.

It was Christmas break at Grand Valley State University in Grand Rapids. It was the year 2011, and I was attending classes and living in an off-campus apartment. Motivated by movies such as *The Matrix* and *Inception,* which I would watch alone at night, my thoughts began to race. They raced with ideas about reality, philosophy, subatomic physics, astronomy, and theoretical physics. I was manic as it is known by psychiatrists and mental health professionals. As my thoughts raced for hours at a time, they fell on some pretty interesting and radical notions. "Follow the logic," I would tell myself as my thoughts formed trains of reason and logic and landed on conclusions like what a black hole really is, and how the Universe was made—that is, how it actually started. Even a theory on how to make jumps in time into the past and future. My mind felt as if it was becoming acute in visual focus and mental processes to the point that I would become dizzy by looking at complex images. I was dabbling in ideas about reality and existence that had only been previously suggested in movies like *Inception* and *The Matrix*. I learned very much during this time in my life, and it was then that I had the idea to write a book. In the time between then and now, I have only gained knowledge, intelligence, and wisdom as it pertains to this book. The title? I'll let you decide; however, I will call it, *My Thoughts Exactly*. Take your time, read it at your leisure, and read at your own risk. Please enjoy. Thank you.

Introduction

Welcome to this book. As I open, I would like you to keep this quote in mind. An important bit of information. It is said in *The Kybalion Hermetic Philosphy*, that "The lips of wisdom are closed, except to the ears of understanding." And so I begin. You might be asking yourself, What is this book about? Well, I would like that to remain a mystery. A mystery for you to solve even after you read the book. Think of it as a riddle to the explanation.

It's all about entertaining the reader, right? My answer is yes. Above all else, I hope that this book leaves you entertained and curious. Of all the answers this book will hold, I would like it to end you with more questions than you have now. After all, they say that the best stories leave you wondering. I want this book to be interesting, entertaining, mysterious, educational, and exciting. I want time to fly by as you read it on an airplane. I want it to captivate you as you read the day away at a lake house. I want it to be your perfect bedtime story. I want it on coffee tables and nightstands. I want it to be seen in cafes, libraries, and bookshops. I want this because I've always been someone who finds fulfillment in teaching and sharing. I hope you learn a great deal from this read. And without further ado, let's get into it. That is, if I have captivated your interest as it is the job of any good introduction.

Chapter 1

The Beginning

Let's start with the beginning. Not of this book but of it all. How everything came to be. How the Universe itself began as I understand it. The Universe began as an idea; an idea is the knowledge that something is possible. This inception is trillions of years old. The idea (the creation) came about with knowledge and wisdom from universes passed. The idea was simple. It was a question. The question was: What if? Moreover, the idea asks, "If I make a universe with this logic, what will its destiny be? What will its fate be?" You see, the creation knows how to make a universe because it has made countless before, and it has learned from them knowledge and wisdom. As the idea grows and becomes concentrated, it creates fine matter or energy. What is energy? Well, that is a question that has yet to be correctly answered by scientists and physicists on Earth, or even by some of the most advanced beings that live beyond the stars of our night sky. Energy is the first thing to come from the creation. The creation being the idea that is the Universe. As the fine matter (energy) grows, it starts to rotate. It rotates and it pulsated. Once a certain amount of energy has been concentrated, the creation has another idea. The idea to create another dimension. The dimension of matter. The energy then finds a singularity where it will spawn the matter. A singularity, what is it? A singularity is simply a point in space and time, and so, space and time begin. The matter is then shot into three different directional vectors, creating the three spatial dimensions of the Universe. The picture below illustrates how the three vectors create a sphere or a three-dimensional universe; all projected from the same singularity. This is, in short, a picture of the birth of space and matter in our Universe.

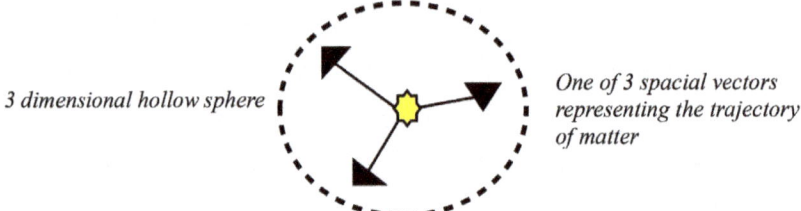

3 dimensional hollow sphere

One of 3 spacial vectors representing the trajectory of matter

It is commonly known in theoretical physics that space and time are but one and stand as the fabric of our Universe. To my knowledge, this is true. The theory of relativity, it is called. Although space and time are woven together into a fabric, they exist as two independent dimensions. Dimensions of our Universe. Matter, being its own dimension as well, affects space and time through gravity. It warps space and time like the structure of a spider web, as shown in this image.

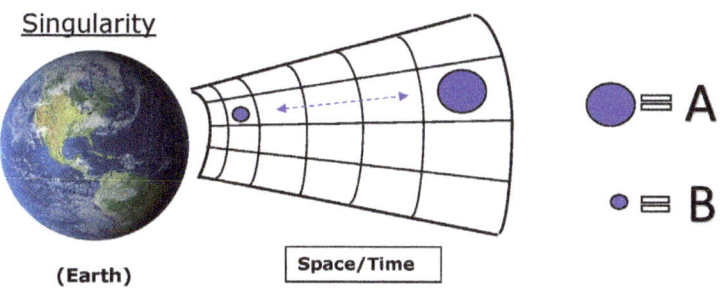

The closer you get to a body of matter, the more you are affected by the warping of space or time. Even matter itself is spatially affected by relativity. For example, in the image above is point A, representing a moon or planet or what you have; maybe even an atom or a body of matter. Point B represents point A as it is affected by the gravity of Earth. As point A gets closer to the planet and becomes point B, its size within space becomes relatively smaller. It shrinks as does space or time. Matter is bound by space.

With this knowledge of space and time relativity, we can come to a logical conclusion as to what happens to matter when it is drawn into the event horizon of the mysterious cosmic gravitational force known as a black hole. Black holes are born from large stars that run out of fuel and explode. This occurrence is known as a *supernova*. Depending on the physics of a supernova, the core of the exploding star may collapse due to an extreme amount of dense matter and gravity. This creates a black hole. It is basically an astronomic vacuum with a possible infinite capacity that sucks matter into itself, into the singularity at its core. It is theorized by our planet's leading cosmologists that matter is pulled apart and destroyed when it enters a black hole and that the stars and planets that get pulled in are colliding and fusing together through heat, similar to what happens when you pack a snowball. However, that theory is not necessarily true. They fail to include the details of their very own knowledge of space and time relativity. The correct answer as to what happens when matter is pulled into a black hole is that almost nothing really changes. This is because matter is bound by space. You see, from a distance, it may seem like stars and planets that are pulled into a black hole are colliding. But in fact, two stars that are a light-year apart will always be a light-year apart no matter how far they go into a black hole. The fabric of space/time would make the size of these stars relatively smaller, including the hypothetical ruler that would measure the distance between. The units of measurement would shrink within the space, along with the stars themselves. These stars would never collide within a black hole because as they would get relatively smaller within their space, so would everything within that space, including the hypothetical ruler that measures the distance between them.

Thereafter, we have the dimension of time. Space or time is like the surface of an ocean. As a metaphor, the big bang being the start of the Universe is that of the wind along the ocean's surface. Wind generates waves and so does the energy of the big bang, relevant to the fabric of space and time. It is these waves that allow for the possibility of time travel into the past. In theory, it can be done by jumping from crest to crest of these waves. Time travel into the future has recently been theorized by physicists like Stephen Hawking. For the first time in known history, we have discovered the possibility of time travel into the future. We have already seen it ourselves in orbital satellites, as well as in the collision of atoms at the speed of light. It's simple. Because of space and time relativity, it is known that our satellites orbiting Earth experience a slightly slower relative time than we do of the surface of Earth. When I say "slightly," I mean a billionth of a second which are known as nanoseconds. Nevertheless, the engineers behind the construction of our satellite's clocks have to account for that time difference. This is actually a modern-day example of actual time travel into the future. How? Well, once again, it's simple. Molecules have what is known in chemistry as a half-life. It is the time in which they break down; how long it takes the aging process to have its effect.

Our satellites experience a slower time than is on Earth, therefore, they travel into the future by returning to Earth younger than let's say, satellites stationed on Earth and waiting to be launched into space. The satellites returning to Earth have lived in time that are nanoseconds slower, relative to the time on the ground. The longer they stay in space, the larger the time dilation becomes. By the time they do return to Earth, they could literally be minutes and possibly hours younger than all matter on the ground, 59.999999999 seconds for our satellites equal sixty seconds for us. Take the same concept and apply it to a spaceship full of astronauts in orbit around a giant black hole with a force of gravity far greater than is Earth's, or astronauts spending time in free space, far from any significant forces of gravity, and you have the potential for time travel years into the future. In fact, the gravity of a black hole is so strong that even photons traveling at the speed of light cannot escape it, that is why they appear black to the eye. They suck in light like a vacuum as well as stars, planets, moons, nebula, asteroids, and comets. Everything gets sucked in due to their immense gravity like pouring liquid into a funnel or watching your bathroom or kitchen sink drain. Forgive me if any of this has been redundant. Furthermore, I digress. The universal speed limit. The speed of light. It gives us another possibility of time travel into the future. Just as gravity affects time, so does the speed of light.

The Hadron Collider, a giant particle accelerator in Switzerland, exists as a track or giant tunnel, about the size of a subway system. Miles around, it sends particles like protons around and around as they approach the speed of light. Subatomic physicists discovered through tests that as particles near the speed, their half-life becomes longer; this is because of the universal speed limit. What happens is that unknown forces that make up the laws of our Universe will actually slow down time itself for matter with enough energy to break the speed of light. It's similar to a governor on a motor. They discovered that as the particles attempted to break the speed of light, they took longer to break down and, thus, lived in a slower time. Taking this idea a step further, we deduce that time travel into the future is possible, given that we figure out how to send people at or near the speed of light. On Earth, if we were able to build a circuit track of some kind, with a vehicle that is capable of taking passengers at the speed of light and with enough energy to surpass the speed of light, time itself would slow down at that point. The passengers would have no awareness of the time difference as this slower relativity in time would affect the motion of everything, including the motion of everything in the brain that perceives time. Enough time spent at this speed would result in a younger vehicle, and younger passengers, relative to those that were not at this speed and living in Earth's standard time relativity. With enough time spent at this speed, the passengers of this speed of light vehicle could be hundreds of years younger than people living on the surface of the Earth in Earth time. In theory, one year spent at this speed could have the relativity of one hundred years on Earth time. The passengers could and would experience only one year of biological aging while people living in Earth time would have experience one hundred years or more of biological aging, in time. This is time travel into the future through simple laws, logic, and physics of the time dimension. Of course, by the time it would take to develop the technology to send people at the speed of light, we will hopefully and most likely have much simpler methods of time travel.

The Universe, the fine matter dimension, the matter dimension, the time dimension, and the space dimension; although it is not known exactly how many dimensions there are to this Universe, at least to the extent of our earthly knowledge, our industry-leading scientists know of one more dimension; they call it the dark matter universe. It was first discovered by American astrophysicists who ran computer simulations in order to learn more about galaxies and their gravity. Before I elaborate, let's make sure you are familiar with what a galaxy is. Simply put, a galaxy is a spinning spiral of billions of stars, each star with planets around it. At the center of every galaxy is what is known as a supermassive black hole and, debatably, a quasar—an extremely dense and sizable collapsed star with the collective gravity of millions of Earths, even millions of stars.

Our galaxy, the Milky Way, is what we see when we look into the sky at night. It is disk-shaped, as are most galaxies of the hundreds of billions known in our Universe. The Milky Way is a type of galaxy called a barred spiral. These galaxies are very rare. It has the figure of a bar near the center, made of millions of stars. Now that you have the rundown of a galaxy, we can get back to the topic at hand; the dark matter dimension, specifically, the computer simulation that discovered it. Astrophysicists ran a complex program that simulated an entire galaxy of matter and the effect gravity would have on itself. What they discovered was that less than half of the matter and its effective gravity was accounted for. The simulations were not behaving as the galaxies in reality do. Approximately 50–70% of the matter via its gravitational effect was missing. They had to figure out an explanation for the missing matter. There had to be more matter in our galaxies, whether it be invisible or even from a different universe, and so, the dark matter dimension became nonfiction. Its matter affects our Universe through gravity and only gravity; that we know of.

It is my personal belief that this very dimension that we call the dark matter world is a universe parallel to ours and is known as the Dal Universe as ours is the Dern; these names I learned from the Pleiadeans of the Dal or the dark matter dimension. These extraterrestrials live in the constellation Taurus of our night sky. More specifically, The Pleiades. Only instead of living in our Universe, the Dern, they live in the dark matter, the Dal. I first learned of them through an alien contact story about a Swiss farmer who had forty years of contacts with these Pleiadeans of the Dal. In the upper-right sector of the constellation Taurus is the Pleiades. Also called The Seven Sisters, it is a group of seven stars very near to each other. The Pleiadeans I speak of live in this location only in the dark matter world, the

Dal. This farmer named Billy Meier did everything from writing books about their teachings, to going on trips into space to other planets, to traveling in time, and even learning telekinesis himself through meditation. It is the single most proven extraterrestrial contact story in known history. It is flawlessly resilient to skepticism. It is as airtight as it gets to speculators. Having studied this story for over seven years, I highly recommend it to those seeking the truth. An amazingly gripping and beautiful story, it covers topics ranging from psychology, to the life of Jesus, to the pyramids, and to meditation. I suggest buying a book by Billy Meier or looking into the Billy Meier contact notes. It's all over the internet.

The absolutum, which is the extent of my knowledge on a matter, leaves us with one final dimension regarding this topic of the dimension of the Universe; the dimension—that is, life. It begins as the dimensions do with an idea. In this case, the idea that life is possible. Flora and Fauna. It is embedded into the DNA or code of the Universe with the intent to evolve as we do. Why does life evolve? Why do we? Because that's the meaning of life according to the Pleiadeans. More precisely, they say that the meaning of life is consciousness-based evolution; to evolve the consciousness of living things. How do we do that? Live as well as you know how. Always progress and learn and, above all, make babies. Your children will be higher evolved than you and your significant other physically, mentally, and spiritually. In this lies the purpose of life similar to the meaning of life. The Pleiadeans say that the purpose of life is to learn and grow in order to contribute knowledge and wisdom to one's spirit. Just as the consciousness evolves, so does the spirit. The human lives its life, so it may learn and grow in love, peace, and harmony. In doing so, the species and all life for that matter evolves. We are to learn from everything that we go through from every experience, whether it be a success or a failure. And in doing this, we gain wisdom whereby wisdom is experience. We are to live in this universe and report the knowledge and wisdom that we gain back to the creation so the creation may learn about its universe through us. In this knowledge about life, we have the answer to why life is difficult sometimes. As living beings, we are agents of God or agents of the creational force behind the birth of the universe. The knowledge and wisdom that we gain and contribute to our spirits is then contributed back to the creation or "God." This is our mission as life forms; therefore, it is not the creation's responsibility to make life easy. It is our job. The best we can do is love and care for one another in this difficult mission that is life, and by doing so, we make our own lives easier while we can maintain strength and maybe even see our lives and the universe as a beautiful and miraculous thing. This brings us to the end of life—death.

    It is just as important as birth and is to life itself as well. I will disclose exactly what happens to one after death. According to the Pleiadeans who are much more highly evolved and knowledgeable in spirit than us, one's spirit is reincarnated often hundreds of years into the future and invests in a more advanced life form. The time in between incarnations is a sleep state for the spirit. When a human becomes brain-dead, that is when irreversible death occurs. When this happens, the spirit leaves the body and joins an atmosphere of spiritual information known as the Akashic records; this is what happens to your spirit after death. The Akashic records exist around planets with spiritually intelligent life.

These records can be accessed through deep meditation and hold millions if not billions of terabytes of information about the past, present, and future. The information can be translated through symbols. I have seen these symbols myself in deep, deep meditation. Written in orange in a domain of red, these symbols flashed before my eyes. Thousands of them. Most advanced humanoid civilizations in our Universe actually use this system of symbols as a universal language and alphabet. These networks of information are planetarily independent, and so, it is not possible to be reincarnated to a different planet than the one on which the life terminated. This is my personal absolutum regarding the life of the spirit after death.

However, what happens to the consciousness after death? Well, the Pleiadeans answered that question too. What they say is that the consciousness leaves the mind at the point of death and joins the timelessness of infinity. It makes perfect sense if you think about it or if you don't think about it. In the absence of time occurs infinity; an almost incomprehensible description of which has no beginning and no end. It exists outside our Universe beyond the bounds of the time dimension. Without time, there is no motion, and without motion, there is no time. So I've learned from Billy Meier's contact notes, this is a very peaceful, calm, and delightful state of consciousness. I describe infinity as frequencies of time increasing and decreasing at exponential rates and without end. What can you take from all of this? I can tell you in my own words. Do not be scared of death. Not only is it an important part of life, but just as important, it is nothing but an experience of heavenly love, peace, tranquility, and bliss. This is what happens to everyone after death no matter how they live their lives.

Different than this, however, is a reality of the afterlife that I experienced firsthand. When I was twenty-four years old, one of the best friends that I have ever had, and one that I lived with for years, died. Following this was the death of my very best friend that I have ever had and one that I grew up with and knew since before I can recall memory. Although I have not really made sense of these things and have not yet come to properly deal with the feelings that are involved, I did have a real experience of the afterlife with an encounter of the angel of the first friend to leave my life, a best friend.

This happened one night while I was sleeping. I felt my heart stop beating and was lightly awakened by it yet not fully awake. A short time after that, maybe thirty seconds, I felt my spirit begin to rise out of my body and my consciousness with it. I was surpassing all of the phases of the afterlife. There were stages. My spirit formed into a ghost but then proceeded to form into an entity, and through a few mysterious phases beyond that, I realized I was transcending into the form of an angel—an angel of heaven. My room was dark; however, everything became bright. As I entered heaven, all my human and earthly stress and problems left me, and to my right was a very bright light, and to my left, I was greeted by the angel of my friend—the one that had died less than a year before this, when I was twenty-four. He appeared as a light-blue and transparent figure. He had the figure of a human, and he had a head although there were no eyes, nose, mouth, ears, or hair. Just a transparent light-blue shape with black lines around the back of the top of his head that went down along his jaw and chin like a chin strap beard. I then looked farther to my left and I saw what looked like a stadium, or an arena, only larger. There were thousands of them—angels. They were all organized into individual groups of about twenty or so, each on their own level or balcony, like a beach hotel, only in the formation of a giant arena. They all had different jobs as angels in this heaven that I was experiencing. They were mostly the same color of the transparent blue and had general figures of a human body. But my friend that greeted me into this arena was with a group on a level or balcony that had these black lines on the figures of their head. His job as an angel, as well as the group that he was with, was and is to save people from darkness. He did not have to tell me this but that is what I learned. He spoke to me, without a physical mouth. He asked me if I wanted to leave my life right now and join him in heaven and work with him as an angel in heaven, an angel of darkness, one that saves people from darkness. What I had to tell him was that I was not ready to die. I had more to live for and more to do on Earth and I told him this. So I left this heavenly realm and returned to my body in my semiconscious state. But then it happened again. I rose into heaven once more and was greeted by him again, and he told me that this was my second chance. He told me that if I wanted to, I could leave my earthly life and join him in heaven. But I couldn't. I knew that I had more to live for and that I had a future in my life on Earth even though my own future was very much unknown to me at this point. So I told him that I couldn't, and with incredible peaceful bliss, I faded away from heaven and fell back into my body and woke up. I will never forget what happened that night.

    I will also never forget what I heard that night. It took what seemed to be about thirty seconds for my spirit to leave my body and rise into heaven, and I heard music. It was amazing. Sometimes people talk about experiences of the afterlife and talk about the music they could hear. What they hear is usually unexplainable to them. However, I have been very proficient in music production for the past fifteen years or more. So what I heard, I could remember when I woke up the next morning. I was actually able to recreate what I heard that night into a thirty-second segment of audio using my music production software. Although I was limited to the capabilities of modern-day music production, I was able to make what was nearly exactly what I heard, and you can experience what I did that night as well. The audio bit and visual that I created that nearly exactly depicts what I experienced, is on YouTube. Just search and find the video titled "What it sounds like to become an angel (an instrumental by Bryton Zaagman)."

It is my belief that we make our own heaven or hell while we live. Those who do terrible things have to live with the choices they make; this can be heaven, or it can be hell. Make your choices wisely, and you will manifest your own heaven. Destiny and fate are two things that I think about often.

If you think mathematically, you probably believe—as most scientists do—that the motion of the Universe was programmed at birth, and that cause and effect rules the present and future. Logic and reason will tell you that destiny is the path of the evolving Universe, and that the big bang was metaphorically the breaking shot at the beginning of a game of billiards. The other view of destiny is that it can be manifested ourselves, and that we determine our own future. As you probably do, I favor the latter. It gives me hope that I am in control of my destiny, and that I can do whatever I want with my life. Of course, as we study more about these two, we learn that both coincide. Sometimes, things happen that can only be described as destiny. Is it the magic of your own manifestation, or is it merely a coincidence? Who really knows? The image below is an illustration of my own original theory of destiny and how it works.

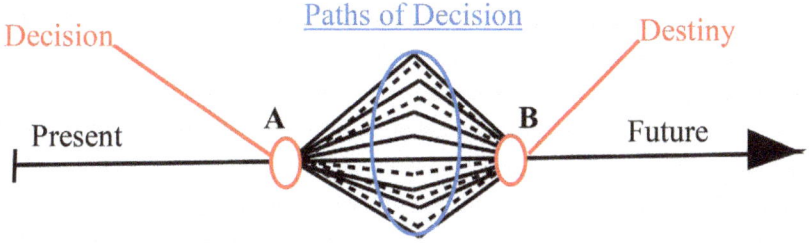

As you can see, the present hits point A and splits into many different paths; however, they then join each other at point B. Point B represents destiny. As point A divides into many paths, each represents different courses of action that we can take, given a fixed present. In going further, point B is an expression of how each path of action, no matter how different, will inevitably lead to the same conclusion or fate in the future. This is a theory that uses aspects from both models of destiny and fate. It is a compromise between the billiard break and the free path, I'll call them. Between destiny and fate, it's all about how you see it. Stay positive, and destiny will bless you; think negatively, and fate will be your demise. Above all else, use your knowledge, wisdom, talent, and intelligence to create your own luck, and in turn, you will be blessed by destiny. Just a reminder, however cliché, luck is where preparation meets opportunity. As I see it, luck is a sign that you are making good choices in life. In my opinion, luck and success are one. If I may, do not view luck as coincidence. There are no coincidences. There is a perfectly reasonable explanation for everything, in so, there is a reason for everything. Think intuitively, and you will unlock the door to luck. Be patient, and you will find the key to the Universe.

    I personally experienced what I can only describe as an example of destiny. It is a fascinating story. One night during my time living in my off-campus apartment at Grand Valley State University, I was invited to a friend's place two complexes down the street from mine for a party. It was a small get-together of about twenty people to watch the Detroit Lions play in an NFL wildcard round of the playoffs. I, being a fan of the Green Bay Packers, brought an iPod touch (state of the art at the time) and a set of Bose IE2 headphones. I was audibly tuned out of the game as I watched and listened to "Sweetness" by Jimmy Eat World. Sipping a beer, I was contemplating many ideas including, and mostly, the concept of destiny. The month or so prior to this was a time in which I was finding success in my undertakings. It was this that led me to ponder the idea of destiny. What was it? How do I harness it?

    Scrolling down my list of artists, I fell upon Rihanna and selected it. Only one song displayed: "Umbrella" featuring Jay-Z. In my opinion, the best song as a whole ever to be made on earth. I selected it and found that the lyrics were sparking my curiosity. They were so beautifully mysterious, almost otherworldly. Jay-Z's introduction reads:

> No clouds in my storms, let it rain. I hydroplane into fame, comin' down with the Dow Jones. When the clouds come, we gone. We Rocafella. We

fly higher than weather, and G5s are better. You know me. In anticipation for precipitation stack chips for the rainy day. Jay, rain man is back with little Ms. Sunshine, Rihanna, where you at?

Such powerful lyrics. I had found a unique and personal understanding of them in the days before. I then cornered my attention on a series of lyrics by Rihanna. Listening to them over and over, I finally found words to the melody. "When the world has took its part, when the world has dealt its cards," she continued, "if the hand is heart, together we'll mend your heart." As I made sense of these words, they began to speak to me, especially at that moment. Ironically, there was a game of poker being played on the table to my four o'clock, Texas hold 'em. Some six or seven people playing as I sat on the sofa delving deeper into the lyrics. "Together we'll mend your heart," Rihanna sang. I wanted my heart mended, but how? "When the world has took its part." It applied so clearly to my life at that moment. It was only a few days after the new year of 2012, and that quote gave itself meaning. "When the world has dealt its cards." It was the perfect hyperbole for my life at that point, literally and figuratively. "If the hand is heart," she sang while the poker table lost a player. "Together we'll mend your heart." Having taken these lyrics to heart myself, I wondered, *How do I get a hand of hearts*? *If I got up right now*, I thought to myself, *would I see a hand of hearts on the table*?

I decided to apply my sense of destiny to this poetic epiphany. *What if I get up and see a hand of clubs*, I asked myself, *or simply a hand at random*?

I used my understanding of destiny and chose to forget about it all and let destiny take its course; that's the only way it would work if I stood aside. Minutes later, having forgotten about these thoughts altogether, I relaxed on the sofa and finished my beer. Without thinking, I stood up and began to walk to the poker table. Step by step, I grew closer to the table lackadaisically, and without the slightest thought of destiny, the song, or the lyrics, I stood at the table and overlooked the action. Just then, the river was being turned. A five-card flush revealed itself. In the middle of the table, it was a flush of hearts. The "hand" was "heart."

I was overcome with exhilaration. I didn't know what to do. I went to the glass door between myself and the back deck, slid it open, and firmly shut it behind me. Acting out of instinct, my right hand lifted and extended all the way into the cold, clear, star-filled night sky. My index finger extended and formed a point. I was pointing to the stars. To this day, I wonder what possessed me to do this. I then lit up a cigarette, a Marlboro special blend menthol, and then enjoyed a tantalizing and delightful conversation with a cute girl—a fellow student at GVSU. It was that night when one of my own mysteries was solved. I will share the clues with you so that one day, you may figure it out yourself. The question is, Where is the lonely island? Your clues lie within this story, and so concludes my story of a personal moment of destiny. Just for your information, the odds of drawing five consective cards of a particular suit from a scrambled deck is that of 1 out of over 2,000, not including the odds of drawing a particular suit within four of them.

This story, and the moment I saw the hand of heart was a candle in the darkness to me. It was a dark time and an even darker night. And I have to say that although it was an amazing moment, those quotes from "Umbrella" and the song itself will always haunt me and have a deep and dark effect on me, probably for the rest of my life. I am by no means saying that song was written for or about me. But in that night, it surely was.

It was a few weeks after that night that I began to understand more about that song. I was paying more attention to the introduction by Jay-Z. There was so much depth about that song in relation to my life at that point and the time after that I cannot even begin to put it into words. That song has taken me from the darkest of dark times and places while being absolutely physically, mentally, and emotionally alone to the brightest of bright days of sunshine, happiness, and enlightenment—and everything in between.

After studying Jay-Z's intro, which was written earlier in this chapter, I discovered that his message in that part was essentially a prophecy. He was prophesizing about the next "Rain Man," so to speak. As he says, "Rain man is back." It was a prophecy for someone, the next great music producer perhaps, to take up the mantel and be the return of rain man, just like he puts it in his lyrics.

He says, "Jay, rain man is back." Now I understood this to be an expression of someone to tell him those same words. Why else would he refer to himself in the third person when he says, "Jay?" What I took from that line was that he wanted the next great music producer to tell him that. He wanted someone to step forward and claim the role of the next rain man and tell him, "Jay, rain man is back."

So that is exactly what I did. Not only was my connection to the depths of that song so significant through all I had been through with it, including my understanding of the Illuminati and seeing Rihanna posing within a triangle itself in the song's music video, but I had actually been working on a phenomenal song over the months prior, working with three young up-and-coming rappers, me being the producer of the instrumental/beat for the song. For that song, and some after, I went by the underground producer alias of Rain Man, and by that, I chose to do what I could to fulfill that prophecy and give the planet the next "Rain Man" as was Jay-Z's prophecy. I even quoted his introductory lines on my social media accounts in an attempt for Jay-Z himself to see it and know that I had taken that place, that mantel, and I made an attempt to fulfill that prophecy.

I even developed my own "Rain Man" outfit which came as a natural expression of the look of my producer alias, and for years in my early twenties, I would wear it when going out to bars and clubs at night in downtown Grand Rapids, Michigan. If you would like to know what that alias outfit looked like, view the picture below, and if you would like to hear the song that I had produced and the one that launched this underground alias of "Rain Man," it is on Youtube.com under the title "1600 (Anthem to the Projections)."

I would like to thank Shawn Carter (Jay-Z) and Robyn Fenty (Rihanna) for helping me and shaping me into the person I am today, for helping me understand that there is incredible freedom in this life and universe, based on the understanding of the science and philosophy behind the Illuminati, which I discovered during this time, and to use those understandings to be free in this world and live by expression of art. And for being there for me in music and spirit while it was raining more than ever.

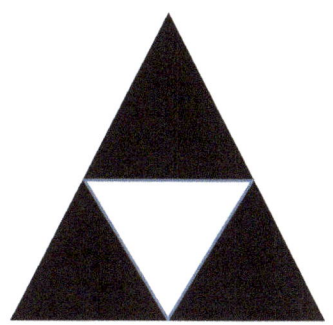

Chapter 2

The Illuminati: The Answers

Now that you have a decent understanding of the start of the Universe and the three dimensions of space as well as the three vectors of matter, we can move on in the topic. This chapter, in proper order to the book's context, will cover my absolutum of how I have made sense of the secret society known as the Illuminati and all of its subjects. We will discuss the triangle, the all-seeing eye, the pyramid on the dollar bill, the Latin on the dollar bill, and the Freemasons—even its relevance to those who "rule the world." Let us begin with the significance of the Illuminati triangle. The word *Illuminati* stems from the words *enlightened* and *illuminated*. The fundamental idea behind the word is that those who are members of this secret society have knowledge and understandings that are considered secrets—of which are only known by its members. As you may have known and or seen, the triangle is a symbolic representation of the Illuminati. It stands for everything that its members believe in and express. The triangle as a shape is the simplest way to implicate the big bang; the birth of the Universe. It does so by implying a singularity. and from there, each corner identifies the three vectors of the creation of our world's three dimensions. In short, the triangle represents the birth of our Universe. Take a look at the image below to better understand this.

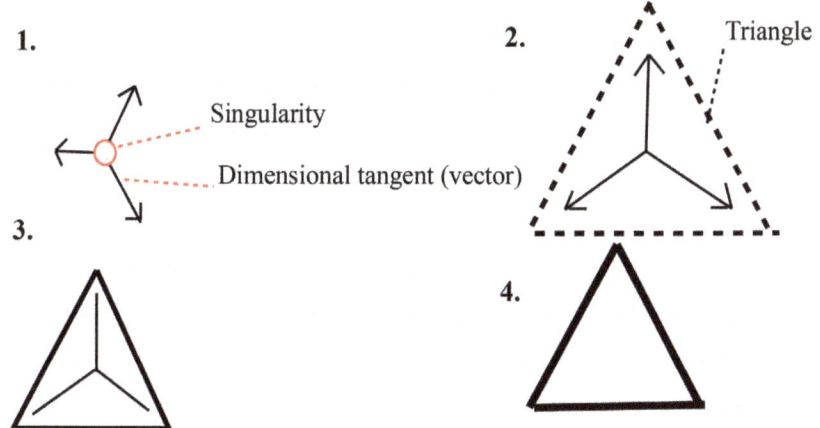

As you can see in the diagram, the triangle represents the birth of the Universe by implying a fixed point or singularity with each of its three corners indicating the three dimensions of space. No matter how you see it, three vectors create three dimensions in space as shown below.

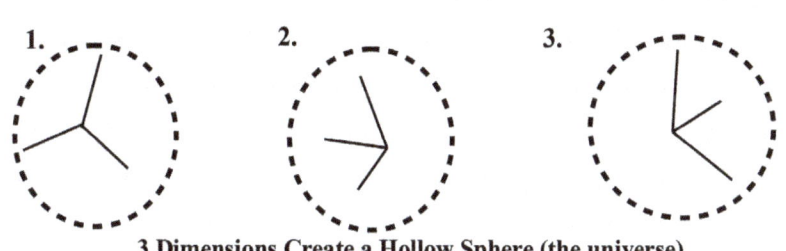

3 Dimensions Create a Hollow Sphere (the universe)

Simple enough, isn't it? I hope so. This brings us to the all-seeing eye and how it plays into the explanation of the Illuminati. When light passes by the iris of the eye, it enters the pupil, the dark circular center of the eye. It does so in many rays or beams. Staring into the mirror and at my own eyes, I noticed how the light met at a sparkling point in the back of the pupil. I analyzed the eye's three dimensions and was reminded of the three vectors of space. Take only three rays of light and follow them to the corners of the triangle, and you have the Illuminati triangle as depicted in the previous illustrations. The image below further explains the relevance of the eye to the Illuminati triangle. Follow any three light rays to the corners of the triangle and you have an expression of three dimensions by three tangents of light.

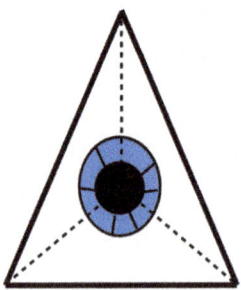

Again, all it is, is a way of most simply implying the birth of the Universe by three tangent vectors. The eye, giving an example of a three-dimensional setting. Well, there you have it. The triangle and the all-seeing eye explained, at least as I see it.

The next step is to accurately understand the significance of the pyramid on the dollar bill. On the first level is a number in roman numerals. It reads "1776" as a number; the year that the Declaration of

Independence was signed and the founding year of the United States of America. The USA was founded on the philosophies of capitalism and free markets as governing principles. The Roman numeral MDCCLXXVI signifies as the foundation of the pyramid. The pyramid itself, topped with the all-seeing eye within a triangle, represents a pyramid scheme in the business of a free and capitalist economy. A pyramid scheme as a firm generates high revenue. It is the most efficient way of making money in a free market. This efficiency is responsible for thriving economies and production, and it is in this, as well as the pyramid structure of corporations, that high revenue is generated, as well as open job opportunities for nearly everyone at every level of education and skill set. It is the essence of capitalism. It often works by using the lower levels to feed the upper levels as shown below.

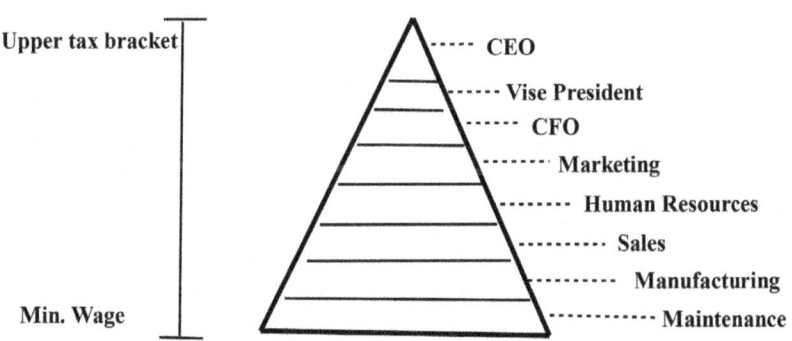

So, in short, the pyramid on the dollar bill depicts and represents a pyramid scheme in business where those in power make large percentages of the profits while creating job opportunities for higher numbers of people at the lower levels of the pyramid format that may be more suited to work in manufacturing or sales. Furthermore of the pyramid, it is surrounded above and below by Latin words *annuit coeptis*, and *e pluribus unum*. Annuit coeptis translates to "he favors our undertakings." This can be interpreted in many ways; however, I believe it refers to a god who favors the new world order. A new nation that, according to the pyramid, is based on the freedom of capitalism and what it is founded on; the principles of the declaration of independence. Annuit coeptis refers to the United States of America in which God favors its undertakings.

E pluribus unum can be translated to "out of many, one." This is in reference to democracy. John Nash, known for his groundbreaking work in sociology and economics, once proved that the best result comes from everyone in a group doing what is best for themselves and the group; working best within the context of a triad or greater. Below is an image that further explains this idea as it applies to macroeconomics. Take a look.

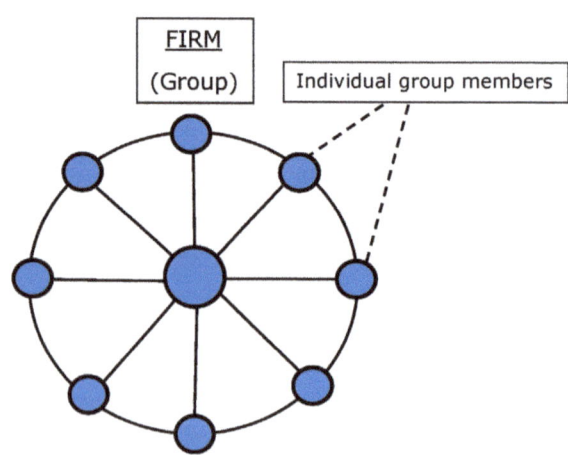

Each member of the group, in this case a firm, contributes to the company by doing what is best for themselves and the group as indicated by the connecting lines. This is a depiction of the logic behind democracy as seen clearly in examples such as the voting process. As millions vote to elect a politician, they become one through a single decision. In short, God favors the undertakings of a democratic nation with a capitalist market wherein it lies a social structure on the brink of pure freedom.

Looking further into John Nash's theory, however, I have found a fatal flaw that possibly renders it redundant. If you consider that the original theory states that the best result comes from everyone in the group doing what is best for themselves, then it is redundant to deem that it is best for themselves "and the group" as Nash's work puts it. Logically, they hold the same definition; mathematically, they are the same. In terms of Algebra, they equal each other and, therefore, the simpler term overrides.

A very real example of this lies in the career and success of NFL quarterback Tom Brady. Tom Brady has won a total of six Super Bowl Championships with the New England Patriots, and as of the year 2021 with the Tampa Bay Buccaneers, he has earned his seventh. Not only has he won a total of seven NFL championships, but he has made his appearance in the Super Bowl (the NFL's Title game) a total of ten times, now with two different teams. His success, I believe, is due to how he conducts himself as a member of his team.

In watching and analyzing Tom Brady's actions during games, it is easy to notice that what he does most of all is his own job. Although he has always been considered a leader of his team, what he does more than anything is let each individual player on the team do their own jobs to the best of their abilities and hold themselves responsible and accountable for their own singular positions. When Tom Brady is off the field while the defense plays for his team, he is usually doing his job—his own job.

I have noticed that when he is off the field, he is not concerned with much of anything, other than bettering himself and what he needs to do when it is his time to go back on the field with the offense. He is usually sitting on the team bench, going over plays and studying himself as to what he should have done differently for better success. His head is down, and he betters himself. He is doing his singular job and minding his own business. Those actions encourage his teammates to do the same, and it has proved itself successful. Herein lies a perfect example of how the best result within a group comes from each member doing what is best for themselves. And in that, it results in a better result for the group as a whole.

This is an example of why John Nash's theory of economics applied to a group as a football team can be considered redundant. Simply put, when each member of a group is doing what is best for themselves, there is a natural reaction that takes place that shows a better result for the group itself. John Nash's theory of economics is implied within the previous theory, making the previous theory hold Nash's within itself and stand true.

On to the last topic of the mysteries of the Illuminati, the Freemasons. The word *mason* means "worker," so Freemason means "free worker," someone who works freely outside of the restraints of a career or a 9:00 to 5:00 job. Many famous artists are considered to be Freemasons; this is because they *are* Freemasons. They work freely and as they wish to make their living. Famous artists like Jay-Z and Rihanna are often seen depicting a triangle with their hands. In the case of these two, I believe that they understand exactly the theories of the Illuminati triangle that are explained earlier in this chapter. I believe many famous and wealthy people have come to these understandings through curiosity. They are more likely to ponder the mysteries of the universe in their private time. Having already found wealth and fame which most people mentally are occupied upon and are in pursuit of.

    They say the Illuminati rules the world. Most people that are viewed by society as members of this "secret society" actually do run the world. In the case of this world, Earth, we are fortunate to have free markets, especially the freedom in the United States. Those that are considered to be members of the Illuminati are oftentimes actually leaders and rulers of the world because their positions become very powerful. They gain millions, and in some cases billions of dollars for themselves; and in addition to incredible fame, they can sometimes hold even more power than governments, particularly when free markets have substantial power over governments to begin with. These people do rule the world as their free work earns them power and money within a relatively free society. The closest thing we've seen to it in millennia; the United States of America.

Chapter 3

How Did We Get Here?

Some people on Earth will spend their entire lives through life and death without enough self-awareness to ask this simple question. The question is this: How did we get here? That is, how did we as humans arrive to this place and time? The place being our planet, Earth. Some people are just too busy with their daily lives to even wonder about this. It takes a very wise and sophisticated mind to have the awareness about the reality of our lives on Earth to even wonder about this. However, there are those here that have an awareness of our place and time in the universe, enough to gaze into the galaxy at night and really wonder about how we actually got here. These people are often very enlightened in the knowledge of astronomy, philosophy, and encounters with aliens. These people are often sophisticated. The word *sophisticated* actually stems from "philosophy" or "philosophers." Philosophers are deep thinking people. They study the big questions in life and try to make sense of our reality.

Firstly, you must know a brief history of our planet and universe. Our scientists have estimated that the universe is approximately fourteen billion years old. However, through my knowledge, I believe it to be much older, closer to thirty billion years old. At least in the history of the age of the existence of what we know of as matter. The dimension of matter is much, much younger than the nonmaterial energy dimension that came before it. Now, the Earth, our planet, is approximately 4.5 billion years old, according to the knowledge of our leading astronomers and astrophysicists. It did not always look like this though. For the first billion years of Earth's existence, it was mostly uninhabitable and was not capable of supporting life. It was in its infant stage, made of mostly rock, lava, and various elements as it gathered mass and density and developed strong gravity. In its young stages, it was gathering matter and growing in size and being pelted with asteroids. Once it became stable and capable of supporting life, we then have two schools of thought. One is, did life on Earth evolve by itself, from single-celled organisms; or the other: was life brought to this planet by other means? The Earth is old enough that it is quite possible that all of the life that we know of today came from single-celled organisms that naturally developed from basic elements and chemicals. It is also possible, however, that life was brought to Earth either by way of asteroids and meteors carrying the building blocks of life or even DNA, or more intriguing, could life have been brought to Earth in a pre-evolved stage by actual alien beings with spacecrafts. The history of life on Earth goes back so far in time that virtually the only way of knowing it all is to research it in the databanks of our galaxies Intergalactic Republic, of which I do not have access to. There is simply too much complex history of life on Earth to be able to determine exactly where specific forms of DNA of all kinds came from and their development over the entire history of Earth other than through our Intergalactic Republic's database of history. Yes, our galaxy has a governing body. The state of our galaxies government, today, is what is called the Intergalactic Republic. It was formed through the joining of four separate galactic governments, within the Milky Way, and was then allied with other galaxies near to ours, hence, the "Intergalactic" Republic. I also do know this: extraterrestrials *have had* major influence on Earth life as we know it, particularly human life. Extraterrestrials played key roles in the formation of all of the human races that we know of today, as well as all of the human races we know of in history. Some small indigenous civilizations in ancient history were evolved from monkeys although there were not many. The reason that we have not found the missing link in the DNA between monkeys and homo sapiens is because there is none, for the most part. It is most likely that the DNA of the races of humans that we have on Earth today was brought here by advanced extraterrestrials. So to the question specifically of how did we get here, the answer is that our DNA was either naturally or synthetically developed and evolved somewhere else in the universe, most likely within our galaxy, and then brought to Earth many times. Many civilizations were brought here to live, and many civilizations were taken from Earth and brought elsewhere for whatever the reason may be. That is the reasoning behind why we have ancient civilizations and cultures that were considered advanced even though they lived tens of thousands of years ago and is also the reasoning behind why we have civilizations and cultures younger than that, that were much more primitive and less evolved than those far older than them.

The most significant factual knowledge I have on the ancient origins of intelligent humanoid life and civilizations on this planet is that there was intelligent humanoid life that lived on the planet Mars. This was most likely millions of years ago, when Mars was in a state where it could support life with an environment and ecosystem to do so. It is to my knowledge that these civilizations on Mars were able to migrate to Earth when Mars was deteriorating in terms of its ability to support life. Yet even with this information, it is still unable to be determined how these intelligent humanoids got to Mars in the first place and asks the same questions about the origins of life on Earth, pertaining to Mars. Even after this migration happened, it is most likely that these civilizations either became extinct on Earth or left Earth via alien spacecrafts. This is my absolutism on the history of the origin of intelligent life on this planet. There is simply too much complex history of life on Earth for anyone to know, outside of the doctors of history of the databases of our Intergalactic Republic.

    This brings us to a similar subject, but in another direction: the future. In the chapter of Revelations from the Bible, the end of the world is prophesized. Now, the Bible itself is very misleading, as most of what was documented came from misunderstandings. The life of Jesus was the most accurately documented information in the entire book of the Bible, and thankfully so, as it is the most important. Yet the origin of Jesus Christ himself was through extraterrestrials. And it is through extraterrestrials that the "apocalypse" will happen. Through my conversations with the people that run the social media account pertaining to the information of the Pleiadean contacts on Earth, I learned that what people refer to as the apocalypse will actually happen. It will not be anything similar to what is pictured in the book of Revelations; however, there will be a large number of people that will leave Earth for another planet and another, better place to live. This will happen sometime in the later half of this millennium, most likely around six to seven hundred years from now. In this time, there will be a separation of Earth humans: those that have found enlightenment and spiritual and biological evolution and those that have not. In this time, it will be necessary to separate these people and to take a small number of the more spiritually advanced people to another planet, one that is more suitable and beneficial for their lives and their own future, where they can live in a place that is more conductive to peace, love, and harmony and their own further growth in all areas of the human life-form. This number of people will be around 150,000, and although they will be given the free will to either stay on Earth or go, there will be a better place offered to them—another planet that is free of the persecution from lesser Earth humans that they have previously lived with here. By this time, the existence of alien life will be a basic standard of education, and it will happen as simply as large alien spaceships landing on Earth and gathering those that are ready to leave. This will be the closest thing to an actual apocalypse concept, at least in the near past or future. In this time, things like the *Star Wars* movies will be looked back upon as nonfiction and basic to a true view of what really happens in galaxies all over the universe, including our own. When these higher enlightened and evolved humans leave Earth by way of extraterrestrials and their technology, it will leave Earth for a negative reaction. There will be less love, peace, and harmony for the time after, as those that have upheld those features will no longer be here. There will be a relapse of sorts where the Earth will experience social destruction, war, and evil. They will have to rebuild, socially and physically.

To the last topic of this discussion, I often find it hard to understand how people can be in such disbelief and doubt of the reality of the real concept of JEDI and SITH knights, even in the Milky Way. In the future, however long it will take, this will become basic knowledge. The JEDI and the SITH are the state of future special operations fighters and mercenaries, nothing more and nothing less. As we evolve in our life-forms in mind, body, and spirit, we will come to achieve forms of super soldiers that are nearly as they are depicted in the *Star Wars* movies. They are not considered soldiers and do not like to be; however, they will be used as such. Just step back for a moment and look into a larger scale of time in the future as well as the past. Most ideas that are considered fiction are always in the process of becoming reality and therefore, nonfiction. This includes the JEDI and SITH from the *Star Wars* movies. Today, we have already begun militarizing our outer space stations. This speaks for itself. It is just one modern-day example that blows the "fiction" right out of the *Star Wars* stories. In addition to this, we have already discovered how to make plasma torches. Plasma is the fourth state of matter following solid, liquid, and gas, and it is what the lightsabers from *Star Wars* are made of. A single unit lightsaber is still hundreds, if not thousands, of years beyond our technology from today; however, we already have fully working plasma torches in the length of what we know of as lightsabers. Although we have not been able to design them to this day as swords for fighting, ours are still powerful enough to dice through all kinds of metal like it is butter, and we use them to cut into locked safes, vault doors, heavily reinforced gates, and things of that nature. Never use your mind to label things as fiction. In fact, there has never been a single solitary thing in our reality, even in advanced science and math that has ever been able to be proved impossible. With a free and open mind comes the knowledge of real truth. Do not be a subject of modern impossibilities. You wouldn't be a flat Earth believer when that was the theory, would you? So don't limit your mind to modern-day theory either. Two hundred years ago, it would be outrageous to believe in the smartphones that we have today, yet today, you are probably either using one to read this; or if not, you may have one in your pocket. The same thing applies to all levels of technology and reality. What is nothing then becomes an idea and, after that, becomes a belief and then becomes reality. There is virtually no such thing as science fiction in this universe, as anything is possible—not only possible, but as a matter of fact, probable and more than likely.

Chapter 4

Witchcraft, the Force, Black Magic, Dreams, Ideas, and Beliefs

They all share a common theology, you may have heard it before. If you believe it, you can do it. It is my opinion that this is a literal law of the Universe. I have spent years dabbling in this philosophy, and I have found it to be true in many ways, and in so, I have come upon a rather interesting quote, an original: "Ideas come from anywhere. Ideas become beliefs. Beliefs become reality."

Let's start with witchcraft. As you will find, this has everything to do with my quote. Contrary to popular belief, witchcraft is a very real and possible thing. It's scary real. What is it? Witchcraft is the manipulation of living and or static things using a point; a wand. It's all in the mind. Witches and wizards use this power by way of the placebo effect. The placebo effect, in its simplest term, is the use of ideas to become reality. Witchcraft, in this case, is nothing more than the application of the placebo effect.

Once again, if you believe it, you can do it. If you believe it, it happens. It becomes reality. How does this apply to witchcraft? Well, witches and wizards use placebo to control living things with the wave of a wand. On humans, for example, the wave of a wand plants the idea in the victim's mind that they can be controlled by it. They believe that they are being controlled, so it becomes a self-fulfilling prophecy all within a single moment. They obey the wand in the thought that it can outpower their own free will when in reality, it is still their own mind in control of their actions. So how do you defend yourself against this? Don't let a wand tell you what to do. You are 100% responsible for your own actions and always will be. You have free will, which is in principle, the byproduct of the law of identity and volition. The law of identity is the confirmed ability to use your brain to formulate what we know as reason. Volition is choice. These two in a mathematical equation equal free will. You have the ability to come to a decision through reason, and you then have the ability to act however you may, based off of that reason or logical conclusion. Witchcraft comes into play when someone can convince you through the point of a wand that they are overpowering your freewill equation. Thought witchcraft was only in fairy tales? I wish. If you ever come across a witch or wizard, remember that you have 100% control of your own body, and nothing can change that.

    As you may have been thinking, the idea of witchcraft may invoke stories from the dark ages or the Salem witch trials of Massachusetts during a younger America. I have been to Salem, Massachusetts, myself. I have seen the burial grounds of nearly one hundred burned, dead, and buried women from that time—women that were literally set fire to and burned to death while alive. Whether these women were innocent or guilty, I have a few beliefs of my own that explain why witchcraft was more prevalent in the dark ages. Hundreds of years ago were times in which medications consisted of natural herbs, small doses of vitamins from plants, and mostly organic treatments for the ill, there were no chemistry labs to synthesize powerful and effective medications for mentally and physically sick people. Because of this, psychosis could hardly be treated leaving people with symptoms like racing thoughts, visual and auditory hallucinations, delusions, and more. This caused many people to have strange, unusual, and radical thoughts and notions. Today, you are hospitalized involuntarily into a psych ward for even something like mild depression. In the dark ages, most of the mentally ill were left with their psychotic symptoms to completely take over their mind. If they had any symptoms as I have had in the past, they probably came upon many ideas that would be considered today as symptoms of schizoaffective disorder or schizophrenia. There were no televisions in the dark ages; no movies, no video games, no cell phones, no tablets, no mp3 players, and no headphones. Most people would finish their workday with wine, beer, or whiskey by a fire leaving them plenty of time to think. I believe that through the placebo effect, many in that time discovered the concept of witchcraft and how it is possible.

Take it from me, anything is possible in this world if you know how to do it. In the dark ages, radical notions were prevalent, much more than today. I think that some who were called witches or wizards found ways to powers like witchcraft through theories of manipulating people's perspectives in application of the placebo effect and the self-fulfilling prophecy.

In the same way that witchcraft is possible, so is the force. Yes, from the Star Wars movies. Instead of using a wand, you use your hand. You can even give people a shove in the opposite direction by throwing your palm out at them. If they believe in your powers, it will become real to them and will control them accordingly. In this, the force becomes real.

Even simpler than this is the power to choke people, just like Darth Vader from the movies. (Disclaimer: this book does not condone this. Do not attempt.) In theory, it starts by putting a hand out in a choking form. It's simple enough. Even the slightest thought that your breathing might be constricted breeds a multiplying placebo effect. When panic sets in, it only contributes to the belief that your breathing is actually being hindered. In this way, the force becomes real. I know and have learned this through personal experience. That is for another chapter through and through. However, one more thing I will disclose about "Force sensitivity" as is depicted in the Star Wars movies is that much of it is no further complex than heightened awareness and senses, even the basic five senses. The Jedi and the Sith, who are known to be very powerful in the Force, are able to hold such powers through nothing more than the heightening of the five basic senses—touch, taste, smell, sight, and hearing. They are known to meditate on a daily basis for hours at a time. Through this, they develop stronger senses and higher awareness.

A natural result of meditation is a clear mind, one that is free from distracting thoughts and mental confusion. A direct result of this is the ability to apply greater attention to one thing at a time, for example, one or more of the five senses. I am not saying that the Jedi and the Sith are not aware of six or more senses. What I am saying is that sometimes what is overlooked as mysterious force powers is nothing more than the application and detection through one or more of the five senses.

For example, If a Jedi or Sith were to know that an enemy were approaching from a far distance or even from the outside of a building, it could be through nothing more than the heightened power of hearing or smelling, no more complex than what guard dogs do when they use their senses to detect intruders to one's house and even the ability to hear if someone has taken a step onto their property, even as quiet as walking on a grass yard outside their house.

The Jedi and Sith are very smart, and an example of their intelligence and their awareness, which is their most important attribute, can be seen in their sight. At any given moment, day or night, you may see shadows. Mostly from outside a window at night, or even in your house, or wherever you may be. Normal Earth humans may never think twice about these shadows, but a shadow that is detected by a Jedi or a Sith is a serious alarm and a warning that there may be danger and threats within the area. The *Sith* are of so much stealth, that they operate with awareness of their own shadows, even at night... especially at night. They also wear clothing that is mostly nonreflective black, which blends into darkness and does not reflect light. This means that their figure remains dark even with a flashlight pointed at them. What a shadow is to the Sith is an error in stealth, and enemies that are in the area that make visible shadows have compromised their cover to the intelligence and tactical minds of the Sith.

Furthermore, the Jedi and the Sith can use their sense of smell, sharpened through meditation, to pick up traces of where someone is, has been, or is going. Once again, like we use dogs to follow the scent of fugitives on the run or criminals that are escaping the police. Although I may save more information pertaining to the Jedi and Sith for another book, I can tell you that they are very real and are at work in not only most galaxies in our universe but in our very own, the Milky Way. You could liken their abilities as advanced special operations forces to that of beings with the natural senses of a black jaguar and the mental evolution and intelligence of a highly evolved alien, which they very much are from our point of view from Earth.

On a final note of the topic of the Force, I will tell you in my most basic terms what the Force actually is. The Force is the cause in which the effect is anything and everything that can possibly be conceived. The Force is responsible for the existence of every possible dimension of reality. The Force is light, and it is responsible for reality and existence.

Black Magic, often confused with witchcraft, is its own thing entirely. So what is it? How does it work? Black magic uses manipulation of shadows and darkness to induce fear and, therefore, generate control and power. My theory is that unknown forces like spirits and entities are able to control the darkness of shadows with evil intent. They do this by worshiping evil. Through the worship of evil comes many abilities. None of which, however, have any concrete effect on someone. Black magic is powerless in the fact that it cannot physically touch you. It can only control you through fear. It cannot physically harm you in the sense that it is only a shadow that you see. Its evil intent is to play with your mind. I will tell you that no matter what doctors say, no matter what mental diagnosis you have, and no matter what others think, what you see, hear, feel, smell, and taste is always real. Just because others can't sense it doesn't mean it's not real. It *is* real. By the definition of real, real is what you experience through the known sense, thus, your senses tell you the truth always. It is a part of reality—your reality. Everyone experiences reality in their own unique and different way; that is fact undeniably. How do you defend yourself from black magic? That's tricky. The most effective way is to take antipsychotic medications; this may inhibit you from seeing shadows. Hallucinations are a symptom of many mental disorders. Don't forget, what you see is real even if it can't be known by other people. Don't lose your mental fortitude. I have seen black magic. I have seen it torture a woman in the emergency room. I've never seen anything like it. The way she quietly screamed as if she was being tormented by fear. "What's going on in room 33?" a nurse said. "An exorcism?" Sketchy stuff.

It's a reoccurring theme of this chapter, of this entire book actually. Ideas, dreams, and beliefs. Just a reminder: an idea is the knowledge that something is possible. In the movie *Inception*, starring Leonardo DiCaprio, the character Mal, then Leonardo's wife, wakes up from spending a lifetime in limbo (unreconstructed dream space) with Leonardo's character, Dom Cobb. In limbo, Dom Cobb performs inception on Mal only to have it backfire in their return to reality. Mal, who in the depth of their dream using a dream machine, committed suicide with her husband, Dom Cobb. When they wake from the dream as you do when you die, according to the movie, Mal takes the idea that her reality isn't real to its extent carrying it over into the lucid reality of the waking state. One night, the night of their anniversary, Mal conspires to commit suicide in the woke reality with her husband so that they may return to a further reality beyond their conscious world. As Dom walks into the hotel suite where they spend their anniversaries, Mal is waiting to talk to him from the windowsill of a building across the street.

"Mal, what are you doing?" Dom asks as he leans out the window and onto the ledge.

"Join me," she says to him.

"Were going to talk about this. It's okay...just step back inside...step back inside," Cobb calmly pleads.

"We're going home," she says. "To our real children."

"No, Mal," Cobb says. "If you jump, you're not going to wake up, remember? You're going to die, now come back inside, please. Just step back inside".

"I'm going to jump. And you are coming with me," she persists.

Mal lets a red dress shoe slip from her foot and fall ten stories down to the street below. "Mal," Cobb says in concerned seriousness. "God dammit, don't do this!" he yells sharply. "James and Phillipa [their children] are waiting at home. They're waiting for us!"

"If I jump, they'll take them away anyway," Mal says.

"Honey, what do you mean?" Cobb asks frantically.

"I filed a letter with our attorney explaining how I'm fearful for my safety. How you've threatened to kill me."

Cobb's face cringes. "Why...why would you do this...why?"

"I freed you from the guilt of choosing to leave them. I love you, Dom," She iterates.

"Mal, look at me, please!" he yells in desperation.

Mal stands up from the sill inches further onto the ledge and with a final attempt to save her life yelling, "Sweetheart!" She closes her eyes and steps off of the ledge and falls to her death.

This scene from *Inception* asks a question interesting and powerful. Could it be possible that the reality of your world is a dream? One that you have not woken up from? This has yet to be proven one way or the other by our leading philosophers and scientists. Neither have they made sense of all that is the mystery of the subconscious. When you sleep, the subconscious is very active. What you experience in a dream is a projection of your subconscious. This happens so fast that as you perceive it, it seems like it's creating itself, and that it's uncontrollable. Like the real world. Lucid dreaming is the awareness of the dream state and the conscious control of your subconscious projection. When one performs a lucid dream, anything becomes possible. If you can imagine it, if you can conceive it, you can live it. It is a natural phenomenon of the sleeping mind.

I myself have successfully done it so many times. I once had a dream that took place at my family's beach house in Grand Haven, Michigan, on the eastern shores of Lake Michigan. A beautiful and peaceful place to be, I might add. In my dream, I was relaxing in the sand around sundown when a large helicopter flew overhead. It was the kind with dual propellers. It was carrying a large pallet of metal drums. On them we're the signs for biohazard and radioactivity. It landed on the beach about fifty yards from where I was. Soldiers exited the helicopter and began unloading the drums. I stood and watched, unsure of what to do. Just then, a soldier pointed at me and told me to go inside. I left the beach and quickly went inside to the cottage. I watched through the windows that viewed the water as they began pouring the nuclear waste into the tide. While they did this, soldiers also began going from house to house, breaking in and shooting and killing people who may have seen what was happening. It felt so real. There was nothing to give it away as a dream. The night of this dream was one that I had during a medication withdrawal. One of the side effects of stopping this medication involved very vivid, very real-feeling dreams.

The next thing I remember was being outside with soldiers pointing their machine guns at me. I put my hands up and froze, yet seconds later, they opened fire, emptying their magazines as they shot me from toe to head. I could feel the bullets ripping through my body. As the streams of bullets crossed my waist, ripping through my abdomen and up through my chest, I was waiting for the first round to hit my head. I was completely confident that it was the reality as it was one of the realest-feeling dreams I have ever had.

There I was, with a stream of bullets, seconds from hitting my head. I thought to myself, this is it, this is how I die. Then another thought crossed my mind, I remembered all of the times having watched *The Matrix.* The movie with Keanu Reeves about a computer program responsible for the reality of the human race on Earth. I remembered times during my mid-college experience when I lived by the theory that if I believed in it, then it was possible—even to the extent of defying death as Neo did at the end of the movie. I lived with so little fear at this time even when doing flips off of the ground and driving at speeds of 130 mph, which only begins to describe all of my thoughts and actions at this time in my life.

In the dream, as the stream of bullets went up through my neck and into my head, I knew I could do anything, although still in belief that I was awake and in conscious reality. I took off of the ground in flight as a rocket-propelled grenade chased me. Still in belief that it was all stone-cold reality, I outflew the rocket and returned to the ground; that's when I took my concept of reality one step further. I remember times when I lived in the philosophy that my reality was a dream; that everything was a projection of my subconscious, and that I could control my world by applying the idea of lucid dreaming to my conscious reality. Even to this day, I notice signs that suggest that my reality could be my own projection.

The soldiers raised their guns to fire at me again; that's when I chose to lucid dream and create my own world. With a single thought, the soldiers vanished and the helicopter was gone. It was like it never happened. I strolled back inside to enjoy the rest of my evening. I noticed that my clothes were different. My hair was in a slicked-wet comb-over. I was wearing my classiest outfit, a black collared shirt with a grey hooded button-up, a black wool jacket over that, and grey slacks with a black leather belt and silver metal buckle with black leather dress shoes and a silver and black Rolex watch on my left wrist. I was straight out of the movie *Inception*. I walked inside having full control of my world— walked into the kitchen where I was met by a room full of people; they were superheroes with the attire to fit it. I thought they were all "whack" for wearing superhero suits. I was more powerful than them. I was a dreamer, then I woke up. At no point during this dream was I aware that I was dreaming, and yet, I was lucid dreaming.

One thought that I have contemplated and that contradicts the idea that your universe is a dream is that everyone is the same. If you have a mind and a body and are conscious with at least five senses, and you consider that other people look just as you do, then it is only logical to assume that they are conscious as well, and that you are at least sharing the same dream space. Unless, of course, you believe that they are projections. By all definitions of *reality* and *real* as terms, it is impossible to decipher a dream from reality. Or should I say, reality from a dream.

Believing in something is a powerful thing. Everything in history has only been done because we believe that we could do it. Beliefs become reality. With the effect of evolution comes more advanced and enlightened minds. Take the sport of halfpipe snowboarding for example. The best snowboarder in the world, Shawn White, pulls off amazing aerials in the halfpipe as well as the slopestyle course. Twenty years ago, a double cork 1080 in a professional halfpipe was considered impossible as magic, but when his sponsor, Red Bull, built a private halfpipe with a practice foam pit for him, he was given the opportunity to learn new groundbreaking moves. In one day, he successfully landed the double cork 1080, front side and backside, as well as the double McTwist, debuting them at the Winter Olympics where he won gold.

He set the bar high with three new tricks that had never been done before. Because it was made known that these maneuvers were possible, it became a belief. It started with an idea by Red Bull that these stunts were possible; an idea passed on to Shawn White. When he started making the full rotations into the foam pit, it became a belief. He believed that he could land it in the halfpipe, and because of this, he was able to do so, and thus, the idea became a belief, and belief became reality. Then after, many others started landing these tricks, and it became the standard, not only for Shawn but for all of the leading snowboarders in the industry. As in anything, progression is slow. We take baby steps in the advancement of technology. In the world of athletics, it's baby steps that set standards; baby steps that break records and bust down doors. All the while, it is ideas becoming reality. Thoughts like, *I can run faster*, *I can jump higher*, *I can do this*, or *it is possible*; step by step we progress, and step by step, ideas become reality. Remember this, your reality is the sum of everything you have ever chose to believe and disbelieve.

Chapter 5

Politics Is Ethics

It is part of the Declaration of Independence itself. The belief that as humans, we are born with inalienable rights. Rights that should be protected and respected. The Declaration of Independence emphasizes rights like life, liberty, and the pursuit of happiness. Whatever your political opinions may be, it is important that they can be backed and defined by ideologies and philosophies that can be proven by science or metaphysics. I personally consider myself a libertarian capitalist. I believe in small, if not, no government at all. Governments become corrupt very easily. Mostly because of contradicting and conflicting compromises in fundamental principles that make up the nation's constitution and bill of rights. The bottom line is this: Does the human have inalienable rights upon birth, and if so, what are they? And depending on what they are, does anyone, including a dominant government and/or commander in chief, have the right to violate those individual human rights under any circumstances? Of all the jobs that governments tend to take on as responsibilities, I believe that the most basic jobs of any government are to establish the rights of its citizens and to protect them and to establish an economy for itself and its citizens and to protect and nurture it.

When the government constitution is not philosophically airtight and founded on tangible science, then loopholes show themselves. They cause confusion and years of debate when it comes to passing new political laws and principles. When loopholes make it possible to pass bills, enactments, and laws that contradict a country's beliefs in inalienable rights, then public sectors tend to gain money and power due to special interest.

I believe in a nation governed by its economy; a free, capitalist economy and the private sector. I believe in a nation similar to the early America; the thirteen colonies. In those times, America was an infant, holding true to its declaration having just successfully rebelled against its mother country, England. Time had not had its effect on a civilization that was vulnerable to the slippery slope of politics and forms of government. It's an exponential process: socialism breeds socialism, communism breeds communism, and down the slope it slips to the country's demise. Every great empire in history has died via the slippery slope. Special interest contributes power to the government as their basic foundational principles were compromised and not airtight.

With a larger government comes higher taxes. Larger governments need more capital to fund agencies and agendas that grow and need more and more funding. Taxes negatively affect citizen's work ethic. Production goes down, motivation drops, and so the death of a nation begins as economies drop into financial depression. Using my home country, the United States for example, production took off as the economy grew and thrived off the freedom that was the early US Government. Regulations were minimal and did not stand in the way of firms, allowing them to succeed and grow very quickly. There was no minimum wage, saving companies their profits by paying employees according to the work they do and paying on commission. Minimum wage kills economies, and it has had its effect. Taxes kill economies; we've seen it too. We left Great Britain to escape taxation without representation, and yet, here we are. Over two hundred years later, with a government living off of taxes, our constitution, primarily written by James Madison with influence from people like Thomas Jefferson, Alexander Hamilton, and John Locke, was flawed because of a compromise on basic principles regarding the rights of citizens.

This allowed the government to start taxing and to increase taxes over time. Essentially, with the opportunity to abuse the loopholes in the constitution and the bill of rights, the government could do whatever they wanted as it pertains to taxes, laws, amendments, agencies, and regulations. The slippery slope to communism was taking its toll and has taken its toll with the peak of socialism during the Obama administration.

There is both science and metaphysics behind politics. Politics is ethics, it is how a government treats its citizens. It's the individual rights, if any, granted to the individual at birth. It all starts with beliefs about the Universe. Take a moment to analyze the diagram below, then we will discuss it.

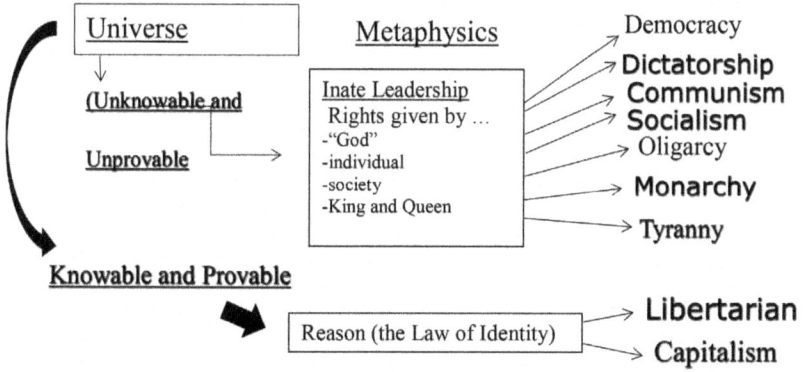

Going Further, politics is all about where you think your rights come from. Let's start with the Universe. If you think our Universe is unknowable and unprovable, then you probably believe that knowledge is innate and comes from a deity like a god; a god that can be communicated with, one that gives you rights based on its own opinion of how people should be treated. The ethics of it leads to a type of government that is ruled by a metaphysical being and has nothing to do with reason which is the law of identity. Such beliefs result in democratic, socialist, and communist leadership. It is self-explanatory, if you believe in a knowable and provable universe, then you probably know it can be done so by using the law of identity. This concept leads to a philosophy that rights can be found using science and rational thinking. The law of identity, also known as reason, was discovered by John Locke, a philosopher who lived hundreds of years in history. It was the reason that was the tool that established the Declaration of Independence and our inalienable rights as individual humans. Unfortunately, Thomas Jefferson compromised on basic principles with Alexander Hamilton, who was a stern communist. The end result? A constitution with loopholes. Two hundred years later, we have laws like seat belt tickets, we have the Drug Enforcement Agency, and we have enforced taxes; all of which are symptoms of socialism and are signs that the country is slipping further down the slippery slope to communism and dictatorship.

Separation of church and state is one of the only remaining blockades on this slippery slope. That is why, it is important that we elect conservative campaigns into office; candidates that still believe in freedom of speech, freedom of the press, and decriminalization of victimless crimes. With all the money we spend on agencies like the DEA, we could fund universal health care and research and technology, leaving citizens the right to take relatively harmless drugs in the privacy of their own homes. Get our economy thriving with more stimulus packages and tax returns. We have seen our economy thriving once again with a president like Donald Trump, a conservative Republican who is very intelligent, knowledgeable, and wise as a CEO. And what is a president but a CEO of an extremely large and complex company, so to speak. When considering the effect of large companies that monopolize and seem to generate an unfair advantage of income to the top jobs within a corporation, it is quite frankly inappropriate to consider the incomes of owners and CEOs as too high or unfair. The income of a single person is no one's business. In most cases, CEOs and owners spend large parts of their lives working very hard to start, develop, and grow their business.

If you want to put it into real perspective, consider that sometimes the people that are looked at as having an unfair financial advantage are the ones that have opened and created job opportunities to the very people doing the judging and complaining about those CEOs salaries. Having done nothing to create that job themselves, some people take their job for granted and forget that those wealthy CEOs have oftentimes created those job opportunities for them in the first place. Case in point: CEOs create jobs. Wealthy people are oftentimes involved in investment banking, which stimulates the economy and *creates* jobs. *Thousands* of jobs. You can thank founders, owners, CEOs, and entrepreneurs of all kinds for most of the jobs that are available to the people and for the state and success of the US economy.

In less than fifty years, this country went from riding horses to walking on the moon. How was this possible?

A healthy economy, one that wasn't being held back by the hand of the government. It was a more prosperous time. In addition to this, I believe we should forget about the national debt. Let us borrow more money and put it into fueling the economy and funding our military. After all, no one can force us to pay it back as long as our military is much stronger and larger and advanced. I also believe that the United States government should develop a new income taxation system. If the citizens of this country could fund the government through a voluntary tax system just like a GoFundMe, I believe that it would be better for everyone. People that live in poverty would have the right to keep 100 percent of their income and be able to more easily support themselves and their families financially while those that make millions of dollars a year would be more likely to take a sense of responsibility and contribute whatever they could afford to fund the government and the important agencies, especially law and order and the military. Although this may seem impractical, I think it would work. With a voluntary tax system, US citizens would individually pitch in whatever they could to fund the important jobs of the government. I believe people have a more generous attitude when they are allowed to donate rather than being forced to. We have seen how a concept like this could be practical. A great example is in disaster relief. When we can unite as one in things like disaster relief, then billions and billions of dollars are generated. I believe a similar system applied to taxes could work. This is the greatest country on the planet. I am proud to be a part of it, and I hope it stays that way for generations.

With freedom comes responsibility. Responsibility is key. When you blame the government for your own mistakes, the government contributes more power to itself by taking on the responsibilities of the individual. They do this by regulating drugs, enforcing seat belts, limiting "texting and driving," and things of that nature—all of which are the responsibility of the individual. As a democracy, if you want freedom, then you better be prepared for the individual responsibilities that go with it. The United States is one of the greatest nations in known earth history. And while I live here, I hope that can continue. As a matter of fact, the intelligent alien life that comes to earth from elsewhere in the Milky Way uses the US first in contacting our planet. The US could be considered as our planet's galactic embassy. They choose the US in most cases because we are the leading nation in our planet, and from an extraterrestrial's point of view, we are the most advanced in morals, righteousness, and peace. Go, USA!

Chapter 6

Astronomy and Religion

"Astronomers had long suspected that there must be planets around other stars in our galaxy, but telescopes were not capable of detecting worlds light-years away. In 1995, astronomers announced the discovery of a planet circling the star 51 Pegasus, fifty light-years from Earth (one light-year equals 5.9 trillion miles or 9.5 trillion km). The 51 Pegasus B, 150 times the mass of Earth, orbits just five million miles (8 million km) from its parent star, just one-sixth Mercury's distance from our sun. This "hot Jupiter" is cloaked by an atmosphere heated to 1800 degrees Fahrenheit or 982 degrees Celsius, but this uninhabitable world is nevertheless a planet, and its discovery proved that other worlds were out there waiting to be found. Other solar exoplanet discoveries followed swiftly. Astronomers have identified more than 3,400 exoplanets. Our Milky Way galaxy harbors an estimated 100 billion planets or more, and at least 1,500 should exist within fifty light-years of Earth. The search is on for Earthlike worlds that may harbor life" (Tom Jones and Ellen Stofan, 101).

This exciting news came over twenty years ago. Today, we have Intel far beyond this. We have the Hubble telescope. Even more exciting than this is the launch of a new telescope; one that will be able to see much farther and more clearly into the Universe. "The James Webb space telescope (JWST) is NASA's successor to the Hubble space telescope. The JWST has a twenty-one feet four (6.5 meters) mirror that will capture infrared light from across the Universe. To be launched in October of 2018, the JWST will orbit nearly a million miles beyond our planet. A large sun shield will chill the telescope's mirror and instruments to below minus 370 degrees Fahrenheit. Among its many astronomical missions, the JWST will observe the atmospheres of earthlike exoplanets, looking for the chemical signatures of life." (Jones and Stofan, 108)

NASA has never been busier. It's a question as old as humankind itself. Are we alone in the Universe? Or, like most astronomers believe, is the Universe teeming with life. Let's look at the facts. "In our Milky Way, planets are the rule rather than the exception: A Kepler study of cool, M-class dwarf stars, which make up 75% of our galaxy's stars, showed each is likely to harbor one or more planets. That means at least 100 billion planets in the Milky Way. Of these, about ten billion should be potentially habitable, earth-size worlds. Adding in K and C-class stars, closer in temperature to our sun, brings the estimate of earth-size planets in the habitable zones around their stars where liquid water can exist on the surface—into the tens of Billions" (Jones and Stofan, 108).

What this means is that there is more than enough opportunity for life to flourish beyond our solar system.

One of my favorite classes I have ever taken was Astronomy in my third semester at Grand Rapids Community College in Michigan. I found the material to be riveting and not just because I was taking Adderall. My world took on a whole new dimension. Up. So much of our lives, we focus on what is directly in front of us: people, work, school, sports, the road, and the sidewalks. So much so that we forget about where we are in the Universe and how much is really out there. If we would just notice the sky at night once in a while, we would be amazed of our own lives and existence. We would be more enlightened.

I remember, as a busy college student, getting home from work or school or the gym on a clear night and stopping to look into the night sky for a few moments just to marvel at its beauty, mystery, and awe. After scoring 100/100 on my Astronomy final, I could locate and identify every constellation in the night sky and even name all of the major stars within the constellation. I would gaze at the moon for a humbling perspective of the reality of outer space: solar systems, galaxies, nebulas, moons, planets, clusters, globular clusters, superclusters, and filaments. I like to think of the night sky as our planet's back yard as I would appreciate the three-dimensional factor of a starry sky. We don't always appreciate the night sky as we should. Next time you look up at night, view it as Earth's three-dimensional backyard. For the most part, brighter stars are closer to us, and dim stars are farther away, although it really depends visually upon the magnitude of the star (how bright a star is). Take into account the fact that the light from stars have been traveling years and years through vast empty space just to enter your eye as photons. Do you ever wonder why stars twinkle? It is the effect that the ozone has on the starlight. It refracts the rays slightly by the movement of the air in our atmosphere, altering the path of the photons. Take time to appreciate the night sky. It could put things in perspective.

My Astronomy professor once said during class that the chance of Earth being the only planet in our known Universe with life is that of everyone on Earth winning the lottery on the same day. Stars are born from nebulas. Nebulas are extremely large clouds of space dust made up of elements like hydrogen, iron, and oxygen. In some cases, these clouds can be light-years across. Nebulas come from supernovae which are exploding stars. These explosions leave behind nebulas. In some cases, these supernovae leave behind neutron stars at the core of the star that exploded. These neutron stars are highly dense, small stars that are white to the eye and look and burn brilliantly like the burning of magnesium. They are so dense that one spoonful of a neutron star has the same weight as the entire city of London, England. Some of these neutron stars become what is known in astronomy and astrophysics as pulsars. They spin very fast and are usually only around fifteen miles in diameter. These pulsars release extreme amounts of electromagnetic radiation.

Solar systems and stars are formed as gravity has a snowball effect on the floating space dust. When millions of tons of dust collect into a sphere, you have a planet. When billions and trillions of tons of dust collect into a mega planet, you have the potential for a star to ignite. When intense pressure builds in the nucleus of these large planets, through high gravitational force, then an atomic reaction occurs, turning the planet into a massive fireball with a core made of plasma. Plasma is like molten lava, only it's so hot in temperature that the atoms break down into a soup of protons, neutrons, and electrons. Stars ignite from the inside and out. When stars get burning, it causes a convection process where plasma floats to the event horizon (the surface), cool down, and then sink back down to the center where the process starts over. Planets then are pulled into the star's orbit by way of its immense gravity, and thus, you have a solar system, a star, and its orbiting planets.

"The Milky Way harbors forty billion earth-size planets capable of supporting life" (Jones and Stofan, 107). The more we learn about our Universe and our galaxy, the more we learn that there is an overwhelming probability that there is not only life but intelligent humanoid life in the galaxy. Not to mention the billions, if not trillions of galaxies in the known Universe. Our nearest galaxy, the Andromeda galaxy, is a neighbor to the Milky Way, and yet it is two million light-years away. I have seen it with my own eyes through a telescope at an observatory in Lowell, Michigan. The telescope was the size of an entire room and probably costs upward of a million dollars. We used infrared flashlights to take notes for a report on the visitation. It was amazing. I saw the Andromeda as a foggy, disk-shaped milky glow. The light from which had traveled for two million years before it entered my eye through the telescope. I saw Neptune as the size of a quarter. I saw Jupiter, our solar system's largest planet, as it looked like a colorful swirly marble. I could see the known storms of the planet as swirling red and white hurricane-like images. It was such a clear sight that I could see three moons in orbit around Jupiter lit white by the light of the sun and in phases just like our moon.

"That's no moon," a classmate commented in reference to the fifth Star Wars movie *The Empire*

Strikes Back. It was truly an incredible experience. I will never forget it.

One frontier we have only recently even theorized is other universes. According to the Pleiadeans of the Dal Universe, there are ten to the fortieth number of universes in a very large universe that we are a part of; that is a ten with forty zeros behind it—a truly astronomical number. This is the extent of their knowledge on the subject and is what they call an absolutum, which means the extent of one's knowledge on a particular subject. According to the race of grey, big-headed extraterrestrials, there are an infinite number of universes, which is a belief of mine as well. This species of alien beings we call "the greys." This is more or less just because of the color of their skin. They have been visiting Earth for centuries, and as far as we know, even longer than that. Their existence is no secret. There have been many government insiders that have stepped forward publicly and disclosed information about them and about one of their ships that crashed in Nevada in the location known as Area 51.

According to the information that was disclosed about this, we were able to study their crashed ship and learn things about their technology. The US government was actually able to reverse engineer some of the technology. This is where we learned of infrared technology, and we were able to design our own infrared lasers as well as learn how to detect and read infrared radiation.

This species of grey aliens which originated from places unknown to me can be dangerous. Most extraterrestrials that we have known to visit our planet are very beautiful, inside and out. It is only natural to assume this, because alien beings that have advanced technology would likely be advanced in biology and spirit, as well.

However, the greys are not so beautiful in terms of emotional understanding and empathetic beauty. This is the case because of simple biology of the brain. The greys have large heads as any science fiction geek knows. Their frontal lobes which are behind the forehead are very large, and it allows them to be very intelligent in the areas of science and math, enough so that they can easily build advanced spacecrafts and travel light years in a single second. The backs of their heads, though, are relatively smaller, and it is in the back of the brain where the processing of emotion and empathy happens. This makes them very smart but not very kind or merciful. You could relate it to modern-day human psychopaths.

They have brains that do not process feelings, emotions, empathy, kindness, and mercy, to the extent that we do. This means that these greys can be dangerous to us. They use us sometimes without much empathy or emotional understanding. They are known to abduct people and use them for scientific experiments and either take them away or leave them for dead. They are psychopaths as we know it. They are not as developed as we are in the emotion centers of the brain. However, they are smart.

It is important to remember that those in the galaxy and beyond that hold the most power in technology and advancement in life are those that are also emotionally and spiritually evolved enough to be of heavenly beauty. The most powerful beings in the universe are the most morally and spiritually "good." They are smart enough to know that the most powerful thing in the universe is love. And in achieving the most power, they have mastered the beauty and art of love.

These are beings of angelic and godly beauty. Even the Sith that are known from the *Star Wars* movies are incredibly loving and caring, contrary to how they are depicted in the movies. They are known to be of evil nature, right? The sense of evil that is tied to them comes from their unending and selfish thirst for power. However, if you think about it, they are quite smart and smart enough to know what the most powerful thing in the universe is as they are in constant search for it. So it only makes sense that in their mastery of power, they have mastered love as it is the most powerful thing in the universe. The Sith are very much loving, caring, passionate, and compassionate life-forms. It is good to know that those that hold the most power in most galaxies are also those that understand love more than anything.

I like to think that gods and angels are real, and I have a quite practical explanation for them. It is an understanding of mine that gods are nothing more or less than actual living humanoid beings. Highly, highly evolved so that they have powers that to us seem impossible. They do this with an extensively evolved mind, body, and spirit, and with the help of advanced technology. Angels are the same, they are advanced in evolution and are incredibly wise, loving, and caring, and they watch over us, whether we know it or not. They are in the form of highly evolved extraterrestrial beings. They are full of love, compassion, empathy, and mercy. Love is the most powerful thing in the Universe, and by it, they perform miracles. These ideas play well into making sense of the stories of the Bible as is one of the Pleiadean teachings. I believe that the Virgin Mary was artificially inseminated by the Angel Gabriel, who was an actual living humanoid extraterrestrial. Gabriel would then be Jesus Christ's biological father. Therefore, Jesus was part Earth human, part ET prophet. He was able to do miraculous things because of a highly evolved mind, body, and spirit, along with advanced knowledge taught to him by his Father. It makes sense. From the Bible, take the word *angel* and *god* and replace it with highly evolved extraterrestrial. Take chariots of fire and lights in the sky and replace it with UFO or spacecraft. Thereby, you have a better understanding; one that makes sense. Thousands of years ago, there was no technology, not even electricity. What do you think they made of ETs landing on the planet? They couldn't explain it, so they called it divine.

For example, on the night of Jesus's birth, the three wise men followed a bright light in the sky. In those days, almost everything that could not be explained was labeled divine; that's how they made sense of things that they didn't understand. Today, an unusually bright light in the sky would be nothing more than a UFO and possibly operated by aliens. That's an example of the progression of how we think in our world today. If not a UFO, then a plane, helicopter, or satellite. A UFO was exactly what the three wise men saw that night. By definition, an unidentified flying object.

In addition, there was little medication for people with psychosis, only herbal and natural substances that had little healing power compared to what we have today. In the times of the Bible, hearing voices and seeing hallucinations were welcomed and praised as contact with god and angels. They had no reason to suspect otherwise. Today, those symptoms would be alarming and would quickly be diagnosed as schizoaffective disorder or schizophrenia. It is my opinion that much of the apparent contact with so-called divine entities were simply symptoms of schizophrenia, including delusions like the idea of god in the first place. Either that or personal encounters with nothing more than actual aliens appearing as divine beings but simply Extra Terrestrials and their advanced minds and technology. Let me further explain this topic with a story.

During World War II, many battles took place in the South Pacific where there are thousands of islands. On a small remote island of one of these lived a tribe of indigenous natives with technology restricted to fire, modest tools, and bamboo huts. One day, a US aircraft flew overhead of the island and dropped a care package with food, utilities, and first aid kits. Years later, another US plane landed on the island and found that the locals had built a model of the plane that gave them the care package. They had built it from wood, and they were worshipping this structure and had developed an entire religion based on it in hopes that it would return with more food and technology.

It is my belief that this is how religions start. It is my belief that this is how Christianity began as well. Stemming from a lack of practical understanding in something that was witnessed, and then there and then labeled divine. It's all just an evolutionary step. According to the Pleiadeans, there are seven evolutionary steps to humankind. The age of reason succeeding the era of religion. After the age of reason, which we are now just beginning, comes the spirit era where understanding of spirit, proof of spirit, and spiritual powers and technology are developed—even higher advanced than the Pleiadeans is the Timmer race from our own Universe; the "Dern" as it is known. The Pleiadeans, who specialize in the spirit, were educated by the Timmers in much of their understandings of the spirit. They teach the importance of meditation for spiritual connectedness and enlightenment. They say that in some cases of Earth humans, the spirit actually leaves the body because it loses interest in the life through the purpose of the physical body. The physical body is a tool that the spirit uses to gain knowledge and wisdom for itself, and if someone chooses to live a life that doesn't contribute knowledge or wisdom, then the spirit will actually abandon the body and leave it to die much faster than those who are spiritually enlightened and are living lives which promote peace, love, and harmony in this world. It is the job of the spirit itself to also evolve and to imbed itself in a reincarnation so it may experience a new and different life in order to gain more knowledge and wisdom. It is our job as humans to always stay learning and experiencing life, as is the purpose of life. When we act on and show love, peace, kindness, mercy, empathy, respect, and happiness, we become part of the meaning of life. I once read a quote that has stuck with me since the year 2011. It read, "Always choose on the basis of love." Not a bad tattoo idea, as I see it. Regarding happiness, I have another quote: "Happiness is making other people happy." Live by these quotes, and you can never go wrong. The Universe will reward you for it in ways you may never understand.

One of my favorite quotes from the Bible is from Jesus Christ. He said, "Whatever you ask for in prayer, believe that you have received it, and it will be yours." A true testament of the law of attraction, which is that if you envision it, it will become reality. These quotes are keys to a wonderful life. They will never misguide you.

While we are on the topic of astronomy and religion, I would like to touch on the concept of gods a little further. Some people that are very religious and dependent on god figures don't even always know exactly what they are believing in. What you should take from this chapter is that there are two different concepts of gods. There is the creation, which is the force of energy that is responsible for the existence of the universe and the birth of it. The creation is the source of energy behind what we know of as the Big Bang.

On the other hand, we have literal existing gods within our universe which are life beings that have evolved to the point where they can actually perform miracles. The Pleiadeans refer to these gods as "IshWish," meaning kings of wisdom. These kings of wisdom are so advanced in evolution of the material and spirit realm that they actually exist more in a spirit and energy state than a material state. Both these categories in concept of God are real, whether you believe in and pray to the creational force behind the universe, or not, or to a closer and perhaps more intimate god that watches over our planet and everything and everyone in it.

The God of the Bible is real. He was real during the times of the Bible, and we were informed of his existence. He is still real and in living form to this day and monitors and watches over the people of Earth. The very same God figure that was documented through the Bible is still living and is still doing his job as God of Earth. The Bible says that this God works in mysterious ways. That is exactly how he works. You could consider him a sort of babysitter over Earth, mostly monitoring and listening to our thoughts and prayers but also stepping in and changing the course of cause and effect through what we know of as miracles.

However, what you must know about our planet's God is that he does let bad things happen. There is a reason for this. Everything that our people go through on this planet is a learning process. Most of the things that happen on this planet are left to our own responsibility as they are learning experiences that are important to us in order for us to naturally learn, grow, evolve, and gain wisdom from. Do not think that there is no help nor intervention in our lives and on our planet from these real-life gods and angels and from the creational energy itself that is responsible for the existence of the universe and even more beautiful, understanding, and loving than our material gods within the universe.

One very clear example of the power and beauty that we have as a valued member of our galaxy and universe is in an example of how extraterrestrial beings once saved the lives of planet Earth. We would not be here and living if it were not for them. If you doubt that there are powers outside our planet with our best intentions in mind, you are wrong.

Being passionate about extraterrestrial life and the influence they have had on our planet and in our lives, I often talk with the people that run the social media account that represents the Pleiadean contact stories through the Earth human Billy Meier. During a conversation, they once told me that it was known that in the year AD 1680, there was a large asteroid headed for Earth with the capability of wiping out all life on Earth. In this year, Earth humans were not only unaware of it but had no way of changing the path of this asteroid or stopping it in any way. That is when a certain civilization of extraterrestrials, most likely from our galaxy, the Milky Way, stepped in to prevent this asteroid from causing another cataclysmic event very similar to what happened in the extinction of the dinosaurs over sixty-five million years ago at the end of the Cretaceous Period.

So when you lose faith and believe that we are helpless and alone in our lives on Earth, remember this: we have a real God. We have the creation. We have real angels. We were given Jesus Christ as an extraterrestrial prophet. And we do most definitely have alien intervention. We are loved and valued as a member of our galaxy in which there are over six million different intelligent humanoid civilizations. We are also a valued member of the entire universe by all intelligent life that is aware of us. How could we not be? Now taking into consideration the figure I gave of six million different intelligent civilizations living in the Milky Way alone, we have to look at the numbers. If you doubt me on that figure, which I have to guess most of you reading this are, then I will give you the numbers. NASA alone is aware of over 100 billion exoplanets in our galaxy. An exoplanet is a planet, just like any in our solar system but defined as a planet within a different solar system in orbit around a star that is not our sun or as it is named, Sol.

After becoming extremely well-educated in astronomy, astrophysics, alien life, theoretical physics, and philosophy, it is my opinion that there are far more planets in our galaxy than there are stars. However, looking at NASA's figure of an estimated 100 billion planets, it is not out of the question to consider that there very well could be six million civilizations in our night sky and mostly independent from one another as they are counted, possibly according to each planet or individual nations on the same planet, like we have on Earth.

When you consider that there are six million individual planets in the Milky Way alone that are home to at least one civilization of intelligent life, it leaves 99,994,000,000 planets to be uninhabited by any intelligent civilization. That number reads ninety nine billion, nine hundred, and ninety four million. And that is according to NASA's estimate of how many planets there are in the Milky Way, which I believe to be a far understatement. So take those odds into consideration if you think that six million is a large number compared to how many exoplanets there are in our galaxy, and for that matter, in our entire universe.

Aliens of all kinds are knowledgeable in the same information given to us through Jesus Christ as it was aliens themselves that gave us Jesus as a prophet to teach us the right way to live. Jesus said, "Treat your neighbor as you would treat yourself." And so why would our galactic and universal neighbors do anything otherwise? These aliens visit us all the time. Look no further than crop circles. They are very mysterious and of popular interest in modern-day culture. Although I would have to say that the majority of them are made by Earth humans, there are some that are not. It is very easy to understand. The crop circles, although few, that are made by extraterrestrials are the signature imprints left by the landing gear of spaceships. Most of the aliens that visit our planet have an agenda or an intent, mostly of good nature. They don't waste their time aggravating conspiracy and hoaxes. They choose crop fields to land in because they obviously need a flat surface to land on, at least the ships that have technology where it is necessary to physically land on the ground.

Crop fields also make good places to land because they can be desolate and miles away from anyone that could observe their landing. I think that the most important reason why aliens of all kinds need to hide their presence is because they know that our societies on Earth are not yet prepared to accept the reality of their existence. They do care for us and they are well aware of our evolutionary state, and because of this, they try to remain out of sight as much as they can until we are ready to have it publicly known that we are not alone in the universe. Given this, there are still thousands of reports of UFOs on a daily basis on this planet.

As a part of astronomy, by fact, I would like to briefly touch on the topic of astrology before ending this chapter. Often confused with astronomy, astrology is the science and theory of how the positions and alignments of the stars and planets have an effect on people's lives, particularly depending on where and when they were born. I was born in March, associating me with the constellation Pisces. The symbol of a Pisces is a fish; the element of Pisces is water; and the ruling planet of Pisces is Neptune, which is a large planet within our solar system that is covered on the surface with very deep water.

Somehow, that element of my astrology has translated into my natural comfort in water. I love to surf and have been surfing since I was in early middle school for the past seventeen years of my life. I have always been fairly comfortable in water behind boats, in deep swimming pools, and tall diving boards. Some people write off astrology as nothing but imaginary and associate it with witches and wizards that look into crystal balls.

Astrology, by fact and through science, is very real. No matter where and when you were born, you will always be affected differently by the frequencies of radiation that are always being given from the stars. Stars and planets are in constant motion, and life is always affected differently by the different kinds of radiation that are put off by stars, depending on where and when you were born and your relation to the position of all celestial objects.

Religion Continued: 666

The truth about the number 666 is this: it does represent the devil, if there ever was such a thing. What it represents more practically, though, is the concept of evil, which I do believe there is. The history and the significance of that number is within the Christian church itself. It is nonnegotiable, as far as I am concerned, with my opinion of the Christian church. Not Christ believers, of which I am one and that I do support, but of what became of the religion of Christianity and the church. It is my belief, and confirmed by the Pleiadeans of the Dal universe, that in the year 666, the Christian church was formed, and with it, the idea of Satan and/or a real devil figure. A lord of evil that lives and rules over a realm that is called hell, which is an afterdeath state. The reality of this is that these ideas of hell and Satan were created by the Christian church in order to instill fear over the people of the church and, therefore, have power over them by threatening them with an imaginary concept of hell itself. For any of you that still subscribe to a conservative Christian-based church, know this: it was the Christian church leaders that created the concept of hell and Satan for the purpose of controlling you through fear of it. The number 666 represents the very thing that you subscribe to. This is nonnegotiable. It will not be debated between you and me, as a writer and reader, as I will simply not support evil of any kind. Do what you will with this information even if it drives you to close this book, but one thing will be made certain: I will not support evil of any kind, no matter where it comes from, even the Christian church. Do not be concerned with hell or Satan, for it is only real if you believe in it, and in believing it, you make it real. There is no such thing as religion—only good versus evil and those that gather to study it. Science versus religion has never nor will ever get anyone anywhere. There is only the truth and it is found through philosophy, meditation, and enlightenment.

Chapter 7

Signs and Symbols

Over the years, I have come to understand many things. None of which rival my expertise in the area of signs and symbols. This chapter will cover my absolutum as it pertains to alchemy, the Illuminati, the Freemasons, peace signs, the Nazis, the Jews, and the triforce.

  Just like witchcraft and black magic, the study of Alchemy holds real transformational powers within its complex symbols. The purpose of alchemy is to unlock the power as a catalyst in order to chemically change elements like metals into each other. For example, an alchemist may use his knowledge of a particular complex symbol, or design, to change gold into copper or magnesium into iron. Every line and shape within an alchemy symbol holds translational intelligence on how to do such things with the understanding of its symbols acting as a catalyst in its own chemistry experiment. Signs and symbols like alphabets can be universally translated into the same concepts, ideas, and understandings. Even each letter within an alphabet can be broken down and traced to a universal symbolic source or cornerstone. It is the proper understandings of the symbols of alchemy that can be used to facilitate an inter-chemical change between two elements. These symbols can be rather complex as they hold great amounts of information. Below is an example of known symbols of alchemy and how they work.

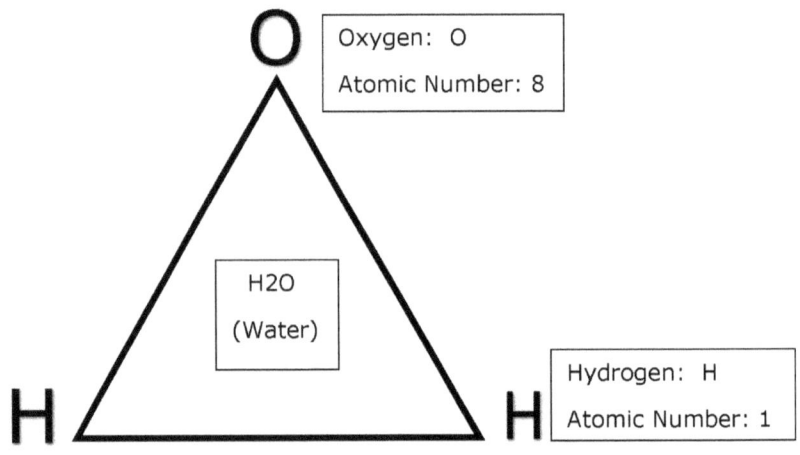

As you can see, this is a very basic example of Alchemy. It involves two parts hydrogen, which is the most abundant element in the Universe, and one part oxygen. As alchemist work, they use signs and symbols to explain chemistry. Instead of a written sequence to explain how chemicals and elements react with each other, they use shapes, signs, and symbols. In this case, a triangle is used to explain the right kind of combination to turn two atoms of hydrogen and one atom of oxygen into water or what is known as H2O. Figuratively, it is more like the wheel of colors and how the combinations turn two completely different-looking colors into completely different itself.

Earlier in this book, we covered the Illuminati and its many aspects. It is founded on the use of a triangle to indicate a singularity and the three spatial dimensions of the Universe. What we didn't cover was its application as a sign in society. As you probably know, triangles are used in public to identify the caution sign within the country's infrastructure. In addition is the application in the console of a car to identify the emergency flashers button. The triangle is used in these instances to help people quickly locate itself. In both cases, it is used to get your attention in the quickest and simplest way. The three corners of the triangle are used to direct visual attention to a point or a space in the middle of its shape. It's no wonder why it's used for important purposes.

In the case of an emergency, like a broken-down car, it is used to get your attention before the other buttons and dials, especially when it involves a potentially stressed and panicked driver and passengers. In the case of street signs, it is used to be the first thing a driver sees as they come into its view, giving them time to slow down and pay attention to a sharp turn or convergence of lanes. This also explains why corporations that use this shape as their brand's logo or insignia tend to find success over those with abstract brand management. Incorporating basic shapes like the triangle, circle, and square will grab consumers' attention and, therefore, develop brand recognition and in turn, more customers and clients. Moreover, incorporating three basic colors like red, yellow, and blue will generate more attention as it is comfortably accepted over blends and combinations of colors. This concept is used in the manufacturing of baby products more than in any other industry. This is for the reason of understanding the root of human psychology and applying it to the products of a business plan.

On to the symbol of the freemasons.

1.

Symbol of the FreeMasons

2.

This design, along with the triangle, is based on the understandings of the Illuminati. The symbol itself, was introduced to Earth by advanced extra-terrestrials. According to them, this insignia of the Freemasons is used in mathematical equations to represent faster than light, speed. These same extra-terrestrials are known to use alchemy to separate gold from seawater; a science that we are only now beginning to theorize and understand. Continuing on the matter of signs and symbols, take a look at the image below which is known as a triforce.

MY THOUGHTS EXACTLY

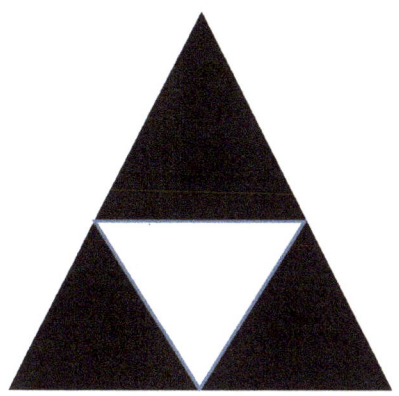

This design is also related to the triangle of the Illuminati, the symbols of the Freemasons, and the triangles used in society for quick attention. This exact symbol is used in accordance with the character Zelda from its stories and video games. It is also used as the logo for the ski producer known as Fischer skis and is labeled on their skiing equipment.

The triforce holds within itself, five triangles. It is sometimes used in relation to the Illuminati; however, it's more well known to be used to represent the word *legend*; hence, the "Legend" of Zelda. The stem of this shape's design lies in extraterrestrial origins. It is greater in power than the standard triangle and also more all-encompassing in its meaning.

The Pleiadeans of the Dal Universe say that the correct peace sign is symmetrically opposite to that of the socially known peace sign. They say that the downward spit means death, contrary to its popular use, and that an upward split means life and peace.

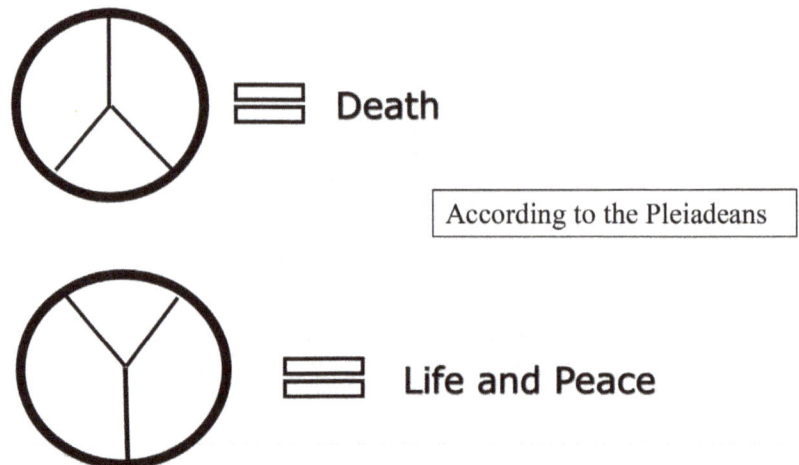

The correct peace sign can then be incorporated into the triforce triangle like so.

Symbols of this nature are most prevalent in the auto industry. With these understandings in mind, it becomes possible to find meaning in even the most basic of designs.

For example, the logo of Mercedes-Benz is that of our culturally known peace sign. It can imply this but can also hold more information, as in reference to the triangle and the three vectors that symbolize the birth of our universe. The logo of Mitsubishi automobiles also holds the same information that could imply the birth of the universe or also a triangle. In the example of Infiniti automobiles, the logo is that of an oval with two lines meeting a point in the center. This symbol is self-explanatory to the Infiniti brand name as it implies infinity by suggesting something like a road that continues straight and becomes visually smaller to no end.

On a darker note, the proper application of deductive reason can shed light on the fundamentals behind the Jewish Star of David and the Nazi swastika. During World War II, Adolf Hitler and the German Nazis were discovering subatomic technology and were making discoveries about atoms and their shapes including the nucleus and the electrons that circle the nucleus at high speeds. To my knowledge, the Star of David, which was given to the Jews by the Nazis, was incepted from a subatomic implication as shown below.

The Atom

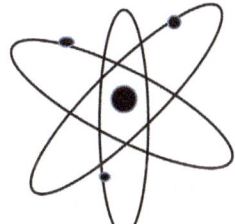

The Jewish Star of David Identified by blue circle

It is quite possible that the Germans, who around the time of World War II were new to atomic research, had settled on the symbol for the Jewish badge by way of drawing an atom in its simplest form and identifying the two crossing triangles through a simple drawing of an atom. The design behind the Nazi Swastika was a direct attempt to signify superiority in the logic behind its inception. As in everything, the subconscious is primarily active and is most influential.

For example, what comes to mind when you see the shape below?

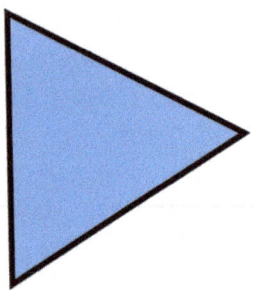

As simple as it may be, there are several meanings that can be translated through this basic symbol or shape. It depicts a play button, a direction arrow, an acute triangle, and more. The more complex the shapes and designs get, the more information they can hold.

This awareness can shed light on even the most complex and confusing written forms of communication. In fact, with the right application of common sense, it becomes easier to translate ancient Egyptian code into another language. Below is an example of how one might make sense of Egyptian writing.

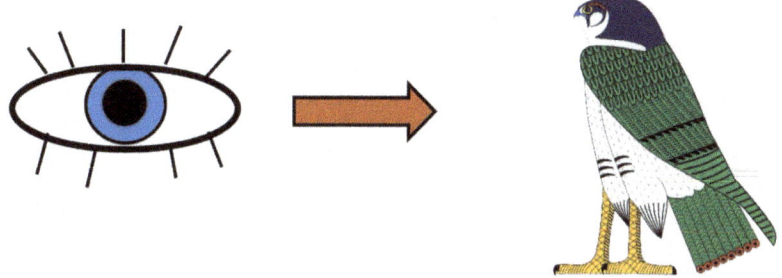

    These known shapes and pictures can be easily translated into basic concepts and ideas. Although Ancient Egyptian writing is often more complex than the pictures above, it can be easily understood using common sense as "I see a bird."

    Even the language called Aurebesh, which is used in the Star Wars movies, can be translated to English by basic recognition of signs and symbols and their meanings. Although it can be known that the Aurebesh written language is based on fiction, the writings can still be translated letter by letter into English. It is my personal opinion that the Aurebesh language was brought to us by extraterrestrials and was then used in the Star Wars movies. It is a very efficient form of writing and understanding. Some words in Aurebesh that only contain as little as four letters can translate to entire sentences in English as well as multiple meanings within each word. It has been my experience that Aurebesh words can be translated with knowledge of the English written language in addition to decipherable hieroglyphics.

    Here below is an example of how one might be able to translate a single four-letter word in Aurebesh into meanings and words in English.

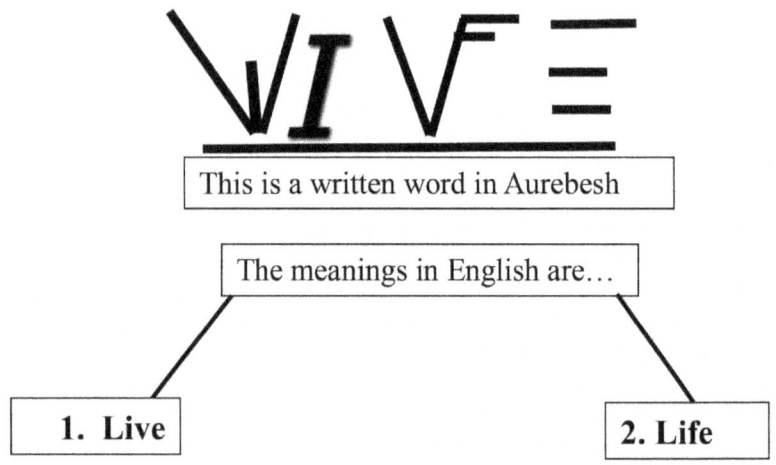

Translating this Aurebesh into English is as simple as studying the lines and letters and putting them together into a cohesive meaning(s). As you can see, the first letter looks like a capital *V* with another line in the middle. The first thing that comes to my mind when I see it is a simple drawing of a plant or "life" as it could mean. The second letter is clearly an italicized *I* just as from the English alphabet. The third letter can be understood as another *V* or an *F* or both at the same time, then you have the three lines which appear as a capital *E* without the vertical line before it. We can take that as an *E* just like it would be used in English, so what we have in this Aurebesh language is a translation to English, meaning "live" and "life." And so, we can put them together and have an English sentence. Live life. And so, we have Aurebesh and the ability to break down its words letter by letter to find meanings in English.

It was a series of events that sparked my interest in Aurebesh as well as the Star Wars movies and books, astronomy, our galaxy the Milky Way, and our place in it all. I am a firm believer in extraterrestrial intelligent life and humanoid beings of all kinds, in our galaxy and all over the Universe. There can and probably are many, many reasons why the general public on Earth has not been informed and enlightened with the knowledge of life on other star systems and our connection to them; mostly ignorance, I believe. Most of us would like to believe that we are not alone in the Universe, yet it is too soon for our planet as a society to be presented with such information that extraterrestrial life is active and in communication and contact with us and our planet.

Chapter 8

Health and Wellness

Regarding the topic of health and wellness, I have quite a bit of knowledge and wisdom.
    There are three parts of the human being that I will be applying my knowledge of health and wellness to the mind, the body, and the spirit. They are all connected and important to one's health and wellness. Let's start with the mind.
    It has been backed up by experiment over experiment.
    Studies in psychology and sociology have shown that our success as humans starts with the mind. It is my belief that at least 90% of success starts with the mind. Mental wellbeing and preparedness with a clear, focused, positive mind is the basis for success in any aspect of life. Belief in the success of one's self is the basis for all success. Like I have alliterated before, beliefs become reality, and a healthy, aware, sharpened mind can facilitate to one's success as their belief in themselves become reality. The self-fulfilling prophecy and the placebo effect work in favor. The simplest thought such as, *I can do this*, is a positive thought and, redundantly,

thoughts become beliefs and beliefs become reality. The best advice for a healthy mind that I can tell you is to meditate, whether it be five minutes or five hours, meditation clears the mind of distracting and busy thoughts leaving the mind with the ability to clearly focus on one thing at a time which often leads to success.

Meditation allows you to become more connected to yourself in feelings and emotions. Your life will take a great change as you learn to meditate and totally clear your mind for the time that you spend in your meditation. It carries over into life, and some people can achieve a constant state of meditation which leads to incredibly positive results. It becomes easier to multitask as well. As a matter of fact, through my own experience, a healthy mind that is free of stress and is cleared of busy and distracting thoughts can allow you to have great success in some of the most daunting and stressful things. For example, public speaking. When your mind is in its healthiest state, not only chemically and biologically but physically (as in its physical shape within the skull), things like public speaking become easy. In this state, you may experience yourself being able to mentally operate and speak to large audiences with nearly no anxiety and with a mind that is performing at its best.

This results in better communication and expression as well as more efficient word use and sentences through intelligence. Intelligence is the ability to draw knowledge from the mind. This kind of mind is not only developed through meditation but is also, in large part, generated through consistent physical exercise. As a last note on the benefits of meditation, I will tell you that all powers in the universe that are within the grasp of human life are able to be obtained through meditation first.

In addition to meditation and positive thinking, exercise is very good for the brain. As you lift weights and run and get exercise in general, your brain gains electrical power which contributes to muscle strength. Muscle mass and muscle density itself is good for the body when it comes to physical health in all parts, but it is the strengthening of the brain's electrical power and signals that contribute to the ability to lift weights and reach the limits of your muscle's endurance. The more electricity that is sent from the brain to the muscle, the more powerful the muscle is as it lifts weight and moves over long-term endurance. As a matter of fact, physical exercise can increase your brain's processing ability and intelligence (the ability to recall information). After exercising the body which, in turn, exercises the brain, one is left with a relaxation phase that leaves the mind very clear and calm without any busy thoughts and sometimes without any at all. This is very healthy for the brain and mind. It is a state of peace and calm and is similar to the effect of meditation. The ability to concentrate and apply the mind to individual things at a given time is a result of the calm and decluttered mind.

When it comes to getting into physical shape and how you want your body to look, I would suggest getting a personal trainer. Although I am not a personal trainer myself, I know most if not more than most do about exercise and developing a physique that you want. Physical image is a touchy subject. Some people just don't have the emotional or physical energy to be in good physical shape. There is a lot of psychology behind the way people look and the way people want to look. It usually starts early in the lives of people. If you were chubby or "fat" as a child, you might develop emotional reactions to that as you get older. Some people who were called fat or made fun of can grow up to have unhealthy eating disorders like anorexia or bulimia. This is extremely unhealthy for the body and emotional mind. It is rather sad, I would say. I have had personal experience of being overweight, as well as becoming borderline anorexic, however, I've never lost the motivation to work out and get exercise through sports like skateboarding, surfing, snowboarding, wakeboarding, skimboarding, basketball, baseball, tennis, and acrobatics, whether it be on trampolines, or in the streets. I am anorexic, as any real doctor or therapist would consider it. I am constantly unhappy with my physique; that is why I have a relentless motivation to work out and get exercise and keep my sugar intake low as well as my fat intake. What makes it so hard for overweight people to lose the fat is because once they lose the fat through exercise, the body naturally wants to replace the fat that was lost. When you are in shape with a healthy fat to muscle content percentage, the less fat you feel like you need to eat because your body only tells you to replace what you've lost, and in the case of someone with high muscle and low-fat content, it is easy to eat healthy with foods that are high in protein, low in fat, low in sodium, and high in natural fruits and vegetables.

It's simple. If you want to have large muscles and have a heavyweight in muscle form, then what you do is lift heavy weights with low repetitions like anywhere from three to at most eight repetitions; this builds large muscles but with low density.

If you want lean muscle and want to have a naturally strong look, you have to lift lighter weights with high repetitions with good lifting and moving form. Running for a half-mile to a mile a few days a week contributes to a smaller, toned look with higher muscle density. As a natural rule, muscle with high density is stronger than softer muscles with low density, although softer and well-rounded muscle shape can result in higher endurance when running or doing activities that involve high endurance.

If you want to be a good fighter, what you want to do is exercise in any and all ways possible, but fights usually don't last for more than a few minutes, so in training to fight, you might want to focus on explosive muscle power and fast sprinting, rather than over long-distance running. The best advice I can give to anyone wanting to become strong and look good with a nice physique is to train the forearms and the calves. When these particular muscles are strong with high endurance and high density, what you have is someone with the strongest body when it comes to all sports, acrobats, and fights. High form high repetition and high form exercise leads to snap speed and higher strength and endurance. Always use good form in a range of motion when exercising. This will quite literally develop stronger muscles with higher endurance as it trains muscles to strengthen through the full range of motion. This trains your muscle's dexterity. Weight training using good form and control through full range of motion develops excellent muscle dexterity. It is often an aspect of strength and athletic ability that is overlooked. However, it is the dexterity of muscles that give them the ability to hold firmly, even while undergoing high nervousness and anxiety. Muscle dexterity is important in competive situations, as it is in these situations that anxiety levels run high. I think that most people do work out and go to the gym to acquire a better appearing physique rather than to be healthier with better physical performance. The most important thing to know about being in shape and looking good is that if you are stronger, then you appear stronger; and if you are faster, then you appear faster. Working smart is just as important as working hard. Even at the gym. Listen to your body and what it wants and needs, and give it so. You will be healthy for it, and although routine exercise is as important as it is, never forget to let yourself get good sleep. An extra hour or two of good sleep is just as healthy for your body and mind as exercise is. So do not sacrifice good sleep for exercise. It sounds obvious and can be almost unpractical, but when you listen to your body and eat when you are hungry and stop when you are full, you will develop a better diet and have a consistently healthy body. As simple as it sounds, be mindful about your diet. Eat when you are hungry, and stop eating when you feel full. This is a simple but very efficient and practical way of getting out of anorexia or obesity.

Now that we have discussed the health and wellness of the mind and body, we go to the health and wellness of the spirit. Some people don't believe in the spirit; some do. I believe that if you pursue an awareness of spirit through an intuitive life with frequent meditation, the spirit will show itself to you. Like we have covered in the book before, the spirit resides in the body, particularly in parts of the brain that not even brain surgeons are aware of yet. The spirit is a part of us in the interest of developing and contributing knowledge and wisdom to itself; that is the will of the spirit.

As we learn more and more about the spirit, we will one day become aware of Midi-chlorians that have only been suggested in the Star Wars movies. Learning about Midi-chlorians starts with the understanding of the word itself. *Midi* is a word often used in music production and is used to describe a piece of equipment like a keyboard to be connected to computers with production software through using the word *midi* as in MIDI connection or a MIDI cable. When you apply that term with *chlorian*, you have a term in which describes something within the chlorophyll of a human cell that is connected and communicates through the body to the spirit with a microscopic life form called a "Midi-chlorian". Hence, Midi-chlorophyll or Midi-chlorian. These micro life forms within the cell are the key communicators between the spirit realm and the physical body. According to the Star Wars movies, they tell us the will of the force, and that they speak to us in their own ways when we are quiet in mind, body, and spirit.

Meditation also helps with stress at an incredible level. It can keep your body in good shape by reducing stress and, therefore, lowering cortisol levels in the body which have negative effects on the body. Cortisol may give men and women more muscle endurance and strength to relieve the body of anxiety; however, it lowers testosterone levels, at least in men. Testosterone is considered to be the master male chemical. It can maintain muscle mass and density and also reduces fat percentages. It also gives men natural energy for productivity in life, and can even reduce sensitivity to pain. Working out in the gym or elsewhere is just as much about generating natural testosterone as it is about damaging and rebuilding muscle. High levels of testosterone in men can maintain muscle mass and density, even in a sedentary life. High levels of free-flowing testosterone in the bloodstream gives men more endurance and strength when it is time to apply physical exercise. Meditate, think positive, exercise, sleep well, and eat well, and it will increase your quality of life and wellbeing to levels you may never have imagined possible.

Chapter 9

Global Warming

Myth, conspiracy, apocalypse—whatever your belief about it may be, the fact is it's happening. Yes, part of it is due to greenhouse gasses; however, many things are contributing to it. And the worst part about it is that the only thing we can do about it is to reduce our carbon dioxide output. Which might as well be a lost cause in our social day and age. People blame it solely on greenhouse gasses, particularly left-wing liberals, if we consider bipartisan politics but in that they are missing the facts. Stopping our planet's carbon dioxide output would take trillions of dollars, and would quite frankly, have the effect on global warming that of throwing a water bottle into a burning building to try to stop the fire. A major contributor to our global rising climate is the concept of planet precession. The science and fact behind this is that our planet (the globe) does not spin on a fixed axis. It may seem as though it does in observance year to year, but in reality, the axis of the polarities that the Earth rotates on actually has a rotation, itself. What happens is a wobbling effect. Yes, the Earth has a wobble on its axis, and about every twenty-six thousand years, it makes a complete 360-degree wobble while still rotating on its axis just like a spinning top wobbling while it spins. This is due to the gravity of the sun and the moon. The axis itself makes a rotational wobble over this amount of time, and the effect is gradual changes in the Earth's climate in terms of global temperatures. These are not the only factors that go into global warming or "climate change." Another factor is that the Earth does not and will never orbit at the same distance to the sun. Our orbit is not perfectly circular. It is somewhat oval-shaped to the degree of difference in distance to the sun of about seven thousand miles every year and closer and farther from the sun. In addition to this is something very important but rarely considered—the temperature of the sun. Our sun, Sol, is in the process of increasing in temperature. This fact speaks for itself in terms of Earth getting warmer. Virtually the only thing that we can do to slow down global warming is to eliminate carbon dioxide output, which traps warm air within our atmosphere.

The simple fact is that there will *always* be climate change. We do not live in a simulation like a video game or computer program. There will always be natural change to our planet. We do not live in a place that will always be the same, which could be designed in a computer program or video game. Everything in our solar system, galaxy, and universe is in constant motion and, therefore, subject to change. Even the celestial equator is in constant change, whether it be a difference of one meter or a thousand kilometers.

The celestial equator is positions and lines given to the planet that designate the sun's location of direct impact through its radiation of heat and photons. Although the Earth is round, there is always a flat enough place on the Earth for the sun to have a direct impact. What is known as the Earth's equator is not always the most direct point of impact from the sun's heat. That is why we have the Tropic of Cancer and the Tropic of Capricorn.

The Earth rotates on an axis. The Earth's axis is tilted to a degree, which causes the sun's strongest impact to change while the Earth orbits the sun and makes a circle around it once every 365 days. This is why there are seasons at all locations around the world. The Earth's equator is simply a median for the sun's up and down target along the face of the Earth. It is the warmest point around the Earth because it has, on average, the most time of direct heat impact.

The Tropic of Cancer and Capricorn then determine the outmost farthest northern or southern points of direct heat impact from the sun. These are degrees of declination on the face of the Earth whether it be farther north or farther south of the Earth's equator. These degrees of declination are correspondent to the Earth's 360-degree shape as it is relatively round.

Our best bet as it pertains to our lives and global warming is to react and adapt. A simple concept that applies to all living things is to react, adapt, and evolve. This is, by no means, the end of us. Where there is a will, there is a way. Necessity is the mother of invention. The people of this planet will certainly have an answer to the problems that global warming creates if these problems ever get serious enough to demand our attention.

Chapter 10

My Outro

This is the final chapter of this book; however, it is only the first book out of many more to come. Consider this book as a single chapter with more to come. As I sign out, what I want to do is get in touch with your deepest thoughts and feelings for a moment. I want to take you to a place in your mind if you can imagine it. We go to a beach. Maybe you have had experiences like this before, and I hope you have because it is a beautiful thing to experienc. If you haven't, then I will tell you how to get there.

We are on a beach. Maybe you are in a room at a beach hotel or resort. If you have a balcony or deck to go out onto, then go out and feel the weather. It is nighttime. The weather is beautiful. The air is in the mid-seventies perhaps. The wind is light, warm, and gently flowing in the direction of land to sea, grooming the small swells as they come into the shore, break gently and quietly in the offshore wind. It is comfortable. The warm land breeze makes for a beautiful and comfortable feeling. These conditions make for the beauty of a dream as there is just enough light from the stars and the moon reflecting off the water, enough to go to the beach without a flashlight. The offshore wind makes the waves quiet and misty as they calmly break and do not make noise, more than a peaceful, gentle, distant rumble.

You have a feeling like you need to escape, like you need to find yourself, find peace, or find someone. You make your way to the ground floor of your hotel, and the night sand is calling you. It has an answer for you, and you need answers. You don't have much, but you don't need much. You don't even have a wristwatch, but that's okay because you don't need it. Your answers are out there along the shore of the water, along the beach, under these stars and moonlight with a beauty that is indescribable because you have only ever felt it before in dreams.

The air is so comfortable, and as it gently flows from land to sea, it grooms the sea and quiets and calms the waves. You are now standing on the beach with the sand still warm from the day before. No one knows you here on this beach and no one knows that you are there. Only you know where you are, and you would trade everything you have or have ever had for this moment, for this time, and for this night.

There is no one else on the beach. It is just you. You walk farther out on the beach and closer to the water and turn right. And you begin taking a walk down this beautiful beach with just enough light from the stars and moon above for you to see where you are going. You don't even know where you are going, but you know it is better than where you have been. You keep walking, and it is just you.

After a while or however long you wish this walk to be, you stop and sit down on the sand. The comfort, peace, and calm of this moment is soothing. It soothes your mind, body, and spirit. For this time that you are sitting on the beach at night, in this setting, you think about anything and everything you wish to, and it helps. It is soothing.

After this, you stand up and slowly and peacefully make your way to the water. You walk just far enough into the tide that the warm water gently washes up and over your feet, and in this moment, the deepest and darkest thoughts and feelings of inner pain and sadness that you have ever contained within yourself come to the surface, and you slowly drop to your knees in the tide, and while looking down, tears begin to fill your eyes. Drop by drop, they fall into the water from your face. Yet in this moment, all that has been from you is silence. Just when you think there is no answer for you, that there are no answers to your life, that you are alone, and that you don't know what to do, just then, you hear a voice. There is someone behind you.

This is anyone that you need it to be of any age, male or female, and of any kind of person that you need it to be. You are on your knees in the tide, and you turn your head and see them. They are wearing white clothes so that their figure is visible in the moonlight amongst the dark of the land behind them. However, the white of their clothes gives their skin and body a dark shade. Not necessarily because of the shade of their skin, but because the white of their clothes hides their appearance.

You slowly stand up and face them, and they speak to you. This is what they say: "Are you all right?"

You don't know what to say because you don't know if you are. This is exactly what you were looking for, this night, on this walk down the beach, and the reason why you are there is the same reason that they are there. You both came to this place, down the beach, to find the same thing, and in finding each other, you have found what you were both searching for this night. You walk closer to this person, and as you do, your pain, loneliness, and sadness evaporates from you and is carried off by the breeze into the open ocean and is gone. You meet this person, whoever you need it to be, and it feels like a miracle.

What happens next is whatever you want or need it to be, although one thing is certain. You sit down with them on the sand and share the loveliest, most beautiful, caring, and soothing conversation of your life. Depending on who this person is, maybe you share a hug or a kiss after a long and deep, quiet, and intimate conversation. If you need the identity of this person to remain a mystery to you this night, then make it so in your mind. You are a perfect match for each other because you have both chosen to go to this beach alone this night for the same reason for the same questions and for the same answers.

The rest of the story is completely what you want it to be, whether you leave them, stay with them, or go with them. Or maybe you are given their contact information, like a phone number so that you may both stay in each other's lives after this night. Make the experience with this person whatever you want or need it to be. This stretch of beach is empty except for you, this person, the stars, the moonlight, the warm sand, and sea. You are completely alone with them, and whether you sit and talk for ten minutes or until the sun rises, it is completely up to you. And you know that this night was the most beautiful night of your life.

As much as I would love to keep writing and writing, I would like to save more of *My Thoughts Exactly* for another time and another book. I dearly hope that this has been an exciting, interesting, tantalizing, entertaining, and informational read for you. As I leave you, I would like to sign out with a few comments. Love is the most powerful thing in the Universe. Spread it through your world. It will return to you in the form of love itself. The love in your life starts with yourself. Spread love to any and everyone you know and meet, and you will experience an incredible life with the return of love to you. Through enlightenment comes the awareness of the beauty and interconnectedness of life and vice versa. Most people in life will sacrifice their health to make money, only to sacrifice that money later to better their health. Always live in the present and cherish it. There is nothing in your life that is worth worrying about, as it accomplishes nothing. Live in the moment, for there only is the moment. There is no past and there is no future. Quite literally, there is only the present. Our reality through life is designed to be lived in the moment, in the present. Living in the past or the future will only lead to stress and problems. A mind, body, and spirit that lives in the moment is achieving life's greatest potential.

In writing this book, there are two specific questions that I would love to answer for you. They are some of my thoughts exactly. Although I want to answers these questions for you, I have no place to put them in the chapters of the book, so I will leave them here. The first question is this: "What is déjà vu?" The phenomenon known as déjà vu is a glimpse, memory, or flashback of a short bit of a dream that you have had. When you experience déjà vu, you feel like you have somehow already lived that moment or have foreseen it coming. This is because you have in your dreams. The subconscious knows a great deal about your future, and when you experience déjà vu. What is happening is that you are recalling that very moment as it was shown to you in a dream, sometimes even years before. Most of your dreams, you do not remember. But when you experience a moment in life that your dreaming mind has shown you before, you experience what is known as déjà vu.

The second question that I feel obligated to answer for you is as follows. It is a popular question, yet it is hard to know the true answer to. You may have heard it before: "If a tree falls down in the forest and no one is around to hear it, does it make a sound?" Now, you might think, *That's ridiculous. Of course, it makes a sound*. My education and experience in life has given me the correct answer to this question. The answer is that the tree does not even exist. If no one is there to witness it fall, then it does not exist. This is because reality is only real upon observation. Everything is energy, and the universe is only real through the interpretation of the stimulus of energy. There is no tree in the forest at all unless someone is there to witness it. If a tree falls down in the forest and no one is there to witness it, not only does it not make a sound, but it has no reality in itself. The universe is not a video game, but the question can be more explained through relating it to one. Ask yourself this, and through this, you have the answer. If there were to be a video game that has a forest with falling trees and you do not turn your video game on and use your character to go to that forest and that tree, does that tree even exist? The universe technically does not take up any size or space or have any much of a reality at all. It is only perceived that it does to the mind. Space, matter, and sound, in this case, are just ideas that are interpreted through the brain when you are there to experience it.

One important message I received recently was a quote from Rob O'Neill, one of our country's best Navy SEALs and a member of SEAL Team Six. Rob O'Neill was the SEAL that killed Osama Bin Laden on the well-known night raid of his house in the Middle East. During a speech he gave, Rob O'Neill said this: "Never give up. Never quit, and you will be fine." I took those words to heart after believing them myself beforehand. Never give up, never quit, and above all, LFA; love first always.

Author's Note

To the reader,
    I want firstly to thank you sincerely for reading this book. I truly wrote it to the best of my ability. I genuinely hope that you found it to be very educational and, just as important, entertaining. What I wish for these remaining pages is to be a kind of reward for completing the read. Because I believe that it was more educational and informative and less entertaining. So what I have for you now is a few stories, like settling down and watching a movie after a long day of work, if you will. These stories are dreams, some of the most vivid and memorable dreams that I have ever had. They are dreams that will remain in my memory.
    These three dreams happened during times in my life where I was withdrawing from psychiatric medications, the side effects from which involved very vivid and real-feeling dreams. Because of that, they will forever be ingrained in my memory. As a matter of fact, they were so vivid and memorable that they are even clearer in my memory than any other given day that I can draw upon from my wake state and daily life. So here it goes, the three most vivid and impacting dreams of my life. They will never be pleasant to remember; however, I do

believe you will find them to be entertaining as even horror movies can be sometimes.

Let me ask you a question. You never really remember the beginning of a dream, do you? You always start out right in the middle of something. So just like that, in this dream, I found myself at a psychiatric hospital or more commonly and more simply known as a mental hospital. Now I have had many visits to mental hospitals, and contrary to the stigma and general opinion of them, they are actually quite nice places to be in. Some of the most beautiful people I have ever met and known were patients or nurses at mental hospitals with me. They are designed similar to an intensive care unit at a general hospital. There are doctors and nurses that take serious care of you and monitor your wellbeing round the clock. The level of care you receive at mental hospitals is actually quite comforting.

The purpose of a mental hospital is to provide you with a safe and comfortable place where you can be assessed and given medications that can help you with mental problems that you may go through like suicidal thoughts, depression, anorexia, anxiety, and more. So in this first dream, I found myself at a mental hospital.

This hospital was a mental facility that was built along the shore of a small lake. This lake was maybe a mile in diameter and had an oval shape. Although this hospital on this lake provided a calm setting, there was an eerie feeling to it. The lake was calm and dark. The depths of the water made it look black. Other than the shores which had small beaches of light and dark sand mixed with grass, plant life as well as down tree branches and earth and forest quickly formed about twenty to forty feet off the water's edge.

So I found myself at this hospital on this lake in this dream. And what I quickly learned about this place was that it had a story. The legend of the hospital on the lake, as it were. The story was spooky and somewhat of a legend or myth. This hospital's spooky story was the legend of the drowned girl. Apparently, everyone that visits the hospital is told of the "spooky legend of the drowned girl" about a girl that had drowned in the lake. And to add to the story, legend had it that her body was never found, even though it was a relatively small lake.

Now, I have never been one to be rattled or frightened about much at all, especially while I am alone. You would think it to be ironic. There is something about me that finds peace and calm when I am in places or situations that have certain levels of what others may call fear. I often enjoy watching scary movies by myself and in the dark.

So I find myself at this hospital in this dream and being aware of the legend of the drowned girl, and the next thing I know is that I decided to go for a walk around the lake. The water was calm with ripples lightly lapping at the sandy and weedy shores. There was just enough clear beach off of the water's edge for me to take a walk around the lake. I found this walk to be very peaceful. The water was very dark, black to the eye, immediately following where the shore dropped off into deeper water.

It was a chilly day, sometime in the autumn season with a light cold breeze that made the remaining leaves on the trees gently sway back and forth. There was a mix of green pine trees and mostly oak trees with green, brown, and orange leaves. The sky was overcast, making the water look black, and the surface was very still and glassy for the most part.

These factors resulted into a somewhat eerie feeling, although quite calm and peaceful to me.

As I was making my walk around this lake, I came to a downed tree that was lying on the shore. I went to step over it, and when I did, I noticed something. There was an old fishing line. It was tied around the downed tree. *Nothing out of the ordinary*, I thought to myself. However, I then noticed that the line went on down the beach about fifty feet or so and was wrapped around another downed tree that was sticking out into the lake a bit. Still nothing unusual, right? You can always find some old fishing line here and there, especially around a wild and overgrown lake.

When I noticed that this line was wrapped around another tree, I also noticed that the line then went out into the lake and into the water. It was then that I got curious. I wondered if there might be a fish or something at the end of this line or that it may have been left there after the hook had been snagged on something. But what I did then, out of curiosity, was to start pulling on this line. I quickly noticed that it was not snagged and that there could be something at the end of the line, maybe a big fish.

I was able to pull this line as it was wound around the other downed tree which was sticking farther out into the water. I kept pulling on it, and I noticed that there was something at the end of it. It was able to be pulled up from the dark depths of this lake, but I didn't know what it was at the end of this line. After a minute or two of pulling the line, I noticed that I was pulling something up from the black of the water and up toward the shallows. I noticed something. There was something at the end of this line. I could see something. I had pulled it up into the shallows and out of the black dropoff; however, the water was still dark, and I couldn't see what it was.

    I stopped pulling the line and I wanted to see what was at the end of the line. So I waded out into the water. After about twenty feet, the water was just over my waist. I walked closer and closer to this thing, whatever it was. As I neared it, something came into view. It was a few feet below the surface, and although the water was black, it had a clarity enough to see what it was. I made my way to it and looked down. And what I saw was a face. A human face. The face of a girl. A young girl. Maybe ten or twelve years old. There was black hair floating around among the water from her head. Her face was pale and white. Her eyes were dead and black. And what I realized was that the legend of the drowned girl of this hospital in this lake was a true story and that what I was looking at was the drowned girl herself.

    You might think that to be a scary thing to see. However, as I usually do, I found it to be a peaceful and calm moment. I was not fearful. As a matter of fact, what I felt more than anything was sorrow. I felt for this little drowned girl. I felt bad for her. I felt emotion with her. Perhaps she drowned by accident while swimming. Or perhaps she was drowned by someone and murdered in that lake. I never made sense of why she was tied to the end of a fishing line. But something about that fact left me with an eerie and darker feel for her story. This poor young girl was dead, and her body had probably been at the bottom of that lake for nearly a year.

    This next dream was just as significant and perhaps an even stronger memory. However, this next dream was very painful and sad, sadder than the one you just read about. This next dream was awful and one that I actually had to talk to a therapist about. Again, I have to remind you that my change in psychiatric medications were giving me very vivid and real feeling dreams.

So in this dream, I found myself in a dark place. I don't know where I was, but I know that it was at night, and I was in a shadowy place next to a building with lights that were too far away which made it hard to see. Now before I go any further, I have to have you know that I have always felt passionate about the US military. I have spent multiple times considering joining the military, whether it be BUD/S (Navy Seal Training), Combat Diver School, or Army Sapper training. I love military men and women. They are so tough, and I respect that, and I have felt the proverbial "calling" toward the US military.

United States military soldiers are nothing short of heroes. They put their lives at risk and fight with blood, sweat, and tears for every ounce of freedom that this country holds.

So as we go further into this dream, it was strange what I found myself doing. In this dark place, at night, under poor lighting, I found myself beating up a young military kid. I have no idea why I would be doing something like that, even in a dream. He was wearing his army clothes, the military-issued camouflage long-sleeved jacket/top and cargo pants. He was also wearing military boots. But for some reason, which really disturbs me, I was beating him up, and he was not putting up much of a fight against me. I was throwing him around, kicking, and punching him. All I could see, really, was the tan and brown of his military clothes, just visible from some kind of floodlight that was maybe a hundred yards away and lighting a parking lot area of the building that we were in the shadow of.

I just kept hurting this young soldier for no reason. I spent about ten or fifteen minutes beating him up, and what I did next was even more disturbing to myself. I grabbed him and carried and dragged him to a nearby trash container. It was near the parking lot that was somewhat lit by a light outside of the building. This trash container was large, almost like a dumpster but maybe a little bigger and deeper. There was a wood pallet in it toward the top. I picked up this young army kid and threw him in the trash container. And as he fell, he broke through the wooden pallet and to the bottom of the container. And I then walked away.

When I got farther away, I stopped because I heard something. This young army soldier was starting to cry. I had left him for dead—beaten, bloody, and bruised—at the bottom of a garbage container at night. I stood there, about a hundred feet away, and heard as he started to cry. He was crying to himself with no expectation of mercy or sympathy from anyone.

I could not stand there. I could not keep walking. I could not leave him like that. I had an overwhelming feeling of sadness and guilt. As I stood there alone, listening to him softly crying to himself, being left for dead at the bottom of this garbage container, I could not stand there and leave him. There was a sadness and guilt that came over me, and I thought to myself, how could I have done that? I thought to myself, *I cannot leave him like that.* It was so sad.

I turned around and went back to the container. I was going to get him back out. So I went back to this garbage container, climbed up, got in, and made my way to him at the bottom to get him out. He was still softly and quietly crying to himself. I felt bad. But what happened next was something even to this day for me is hard to live with, even though it was a dream, albeit a vivid dream. I put my arms around him and began to pick him up to get him out, and as I lifted him, the light of the moon that night graced his face. And to my absolute horror, what I saw was not an army kid. It was not even a boy. It was a girl, a female military woman. A young girl. A female soldier. She was absolutely beautiful. She had medium-length black hair that shined in the moonlight, and her eyes were glassy in the moonlight with tears, and her face was wet with tears.

I could not believe it. I stood there in that trash container, holding her, with the shocking realization that I had done the very last thing I could ever live with myself for doing. I had beaten up a girl. Not just a girl but an army girl. I had nearly killed her.

My emotions were rattled, and I broke out into tears. I could not believe what I had done. I began to cry. There were tears pouring out of my eyes, even more than hers, as I began to feel the pain, sadness, and guilt of what I had done. I felt so bad for her. I was devastated by my own actions and began to bawl and cry with immeasurable sadness for this poor girl. I was crying nearly as hard as I have in my entire life about some of the most difficult things you could possibly imagine, things pertaining to my real life in the wake state as this was a dream.

I was shaking and pouring my emotions of guilt and incredible sadness out over this girl as I was trying to hold her. I then realized that I was the only one left that was crying. She wasn't crying anymore. And in my deepest sadness and sorrow for her, I looked to her face through tear-filled eyes. That beautiful face. That innocent face. It was still in the moonlight. She was so beautiful. And I looked at her, wondering how I could possibly live with myself, and what I saw from her was something that was even more beautiful. She was calm, although she was hurt and had tears on her face. And the beautiful thing was that she was looking into me with a face of forgiveness. She was forgiving me for what I had done to her through the beauty that she was. She knew that I was unaware of her gender. She knew that I thought she was a boy. And although that didn't make it okay, what did make it okay was that she was forgiving me for nearly beating her to death and leaving her at the bottom of a garbage container, leaving her to cry to herself.

I held her tighter and hugged her, and with all the love I could possibly summon, I began to tell her over and over again though my crying, "I'm sorry, I'm sorry. I'm so sorry."

And she forgave me. She was a girl and so, so beautiful. And I had nearly beaten her to death. But through her beauty, she forgave me as the moonlight graced her gorgeous face. And then I awoke in my bed with pools of tears in my eyes.

Both of these dreams are still difficult for me, even months and months after. Every time I see a lake or hear about fishing, it brings me back to that drowned girl and her spooky legend at the mental hospital on the shore. And I still think about that dream of beating up a girl, an army girl, the kind of girl I hold dear to my heart. But that dream in particular brings a sense of beauty, of forgiveness. I was struck by such a sense of love and beauty though her ability to forgive me for that and her understanding that I did not know that she was a girl and that she knew that I would have never done that, especially if I knew that she was a girl. And she was young as well, which makes it even more difficult for me.

The day after that dream, I was in a state of shock, trauma, and awe and left with a touch of grace and beauty from her, so much so to the point that the nurses and staff at the place that I live at noticed it in me and sat me down to talk. And I told them about that dream with my therapist there as well. They tried to tell me that it was just a dream and that I was going to be all right. But it was such a powerful dream of absolute sadness that it was hard to pass off as just a dream. It was so vivid and real. It was incredibly beautiful what I saw in that army girl and her ability to forgive me. But it also was so, so sad.

I have found peace in the devastation of that dream. I don't think I could have if it weren't for the lovely grace, beauty, and mercy in her, enough to forgive me for what I did.

This next dream is the last of the three and will conclude this book. Again, I was dealing with very vivid dreams as side effects of medication changes. It began at a college house. I was in my college days in this dream. I was at an off-campus house of some friends of mine that went to a different college in Michigan. We were having a bit of a party. There were maybe twenty or so people there, girls and boys. In my memory, the party itself went by rather quickly, and I found myself at the house toward the end of the night, around 3:00 a.m. or so. I had met this college girl and was having a great time with her, and I liked her. In a friendly way, she was very cute and fun. But toward the end of the party, she had gone upstairs. I wasn't sure exactly what she was doing up there, but I thought she must be hanging out with some other friends or drinking some more or maybe smoking some marijuana.

I was on the main floor at this point, and people were leaving as it was nearing the end of the party and the night. I had been drinking and was under the influence of alcohol and aware of it, even in my dreaming mind. As everyone left, I was the only guest still there, and it was the college house residents and I. Through some drunk logic, my friends started to talk about their house and that they wanted a better one. So they began to talk, and it led to an idea. What they wanted to do was commit insurance fraud by setting fire to the house in order to get a new house through their insurance. So the next thing I knew, we had containers of gasoline.

Now, through my experience with fireworks of all kinds, gunpowder, flamethrowers, flammable fluids, and things of that nature, I trusted that everyone would leave me responsible for when and where to light this fire. I have experience with the explosiveness of gasoline and how it burns, and my friends know that of me. After pouring gallons of gasoline all over the main floor, they poured a trail flowing out the front door, down the front steps, and to the sidewalk. Being absentminded as these friends were sometimes, as well as being intoxicated, they quickly asked each other in making sure there was no one left inside the house. They shortly brushed over the assumption that the house was empty and safe to light up.

We were at the sidewalk at the end of the gasoline line, and as they were talking about anyone that could still be in the house, I remembered the girl, that super-fun friend I met that night, the cute one. I remembered when she had gone upstairs. But I never saw her come back down and leave. And as I turned to tell my friends I was going to go up and get her, I saw a lit zippo lighter in midair. All that crossed my mind was the name of this girl that I was certain was still upstairs. But it was too late. The zippo lit the gasoline vapor from above, and the trail ignited and ripped into the house with an explosion of flames and intense heat. I looked to the upper floor, and there was a window, a single window, but it wasn't designed to be opened and had wooden crossbeams. I could see the flames coming into that room, and I saw her. She was trying to get out, but she couldn't.

I tried to run into the house and up the stairs, but even the stairs had erupted into intense flames. I tried even letting the flames burn me. But it was too much. I couldn't get to her. I didn't even know what I would have done if I did get to her. The entire main floor and staircase was roaring with blazing flames. I had to watch as the flames consumed the upstairs room, and she fell to the ground from the smoke and the room filled with fire. And I had to say goodbye to her.

And there you have it. Another sad, sad dream about a young girl who went through something devastating to be known by me in these dreams, in these three single dreams that will stay in my memory until the day I die, more than any other dreams that I have ever had in my life. And so story time is over, and so is the last of this book.

I want to thank you again for reading. This book was very important to me, and I was able to express much of myself through it. This is the first official book that I have written, and there will be more to come. Spread the word of this book around to friends and family, if you would like, and if you would be so kind to do so. I will see you again in my next book, the next chapter of my life and of my expression through writing. Remember this: love first always. Live by those three words, and you will never go wrong. Namaste to you all. Thanks.

References

Jones, Thomas D., and Ellen R. Stofan. *National Geographic: The Next Earth.* New York, United States: Time Inc., 2017.

Works Cited Total Information

National Geographic: The Next Earth
produced by National Geographic Partners, LLC
1145 17th Street NW Washington, DC 20036-4688 USA
Copyright 2017 National Geographic Partners, LLC
Text copyright 2017 Ellen R. Stofan and Thomas D. Jones. All rights reserved. National Geographic and Yellow Border Design are trademarks of the National Geographic Society, Used under license. Material in this book is drawn from the following National Geographic book: Planetology, 2008 ISSN 2160-7141
Printed and Distributed by Time Inc. Books 225 Liberty Street New York, NY 10281
To order this or other National Geographic's collectors editions, visit us online at shopping.com/Specialeditions.
Printed in the USA

About the Author

Bryton Zaagman was born on March 14, 1991. The reason he wrote this book, *My Thoughts Exactly*, is that he had acquired a decent amount of knowledge through his life, not only through high school and his attendance at two different colleges, but through his experiences in body, mind, and spirit with his interests and fascinations pertaining to almost every aspect of life. Although he had plans to write a book since early in the year 2012, it wasn't until 2017 that he actually began writing and made a commitment to documenting his thoughts, wisdom, and knowledge. It is his hope that in reading this book, the reader may gain knowledge and understanding about many aspects of life and the Universe, as well as provide answers to some of the least understood and mysterious subjects and concepts that we face today in our world. In this book, Bryton J Zaagman shares his thoughts about the birth of the Universe itself, valuable information about the science of politics, and everything in between.

Bryton is interested in many things and has experience in many things as well. His hobbies include music production for which he produces instrumentals for a record label in Chicago called Astronomic Music Group LLC (which is published by Warner Bros. Records), playing drum sets for a punk or rock band, playing drumline snare drums, surfing, skateboarding, snowboarding, wakeboarding, astronomy, business and business management (as was his focus in college at Grand Rapids Community College and Grand Valley State University), the stock market, brand management, his interests and involvement in extraterrestrials including what are known as the SITH of the Milky Way Galaxy (our galaxy), parkour and trampoline acrobats, car racing, basketball, baseball, football, tennis, getting along with friends and having fun times with them, and as well as getting exercise on a regular basis.

The title of his book *My Thoughts Exactly* applies to this book in a great way. It is quite literally his thoughts exactly of which he has accumulated over his lifetime. In signing off, he would like to express a quote that he made himself and hopes that it will catch on with the people of this great planet. He would like for you to enjoy reading this book, and he would like to remind you, love first always.

Thank you!

www.ingramcontent.com/pod-product-compliance
Lightning Source LLC
Chambersburg PA
CBHW040517220526
45473CB00012B/2888